Market Madness

Market Madness

A Century of Oil Panics, Crises, and Crashes

BLAKE C. CLAYTON

A Council on Foreign Relations Book

OXFORD
UNIVERSITY PRESS

Oxford University Press is a department of the University of Oxford.
It furthers the University's objective of excellence in research, scholarship,
and education by publishing worldwide.

Oxford New York
Auckland Cape Town Dar es Salaam Hong Kong Karachi
Kuala Lumpur Madrid Melbourne Mexico City Nairobi
New Delhi Shanghai Taipei Toronto

With offices in
Argentina Austria Brazil Chile Czech Republic France Greece
Guatemala Hungary Italy Japan Poland Portugal Singapore
South Korea Switzerland Thailand Turkey Ukraine Vietnam

Oxford is a registered trade mark of Oxford University Press
in the UK and certain other countries.

Published in the United States of America by
Oxford University Press
198 Madison Avenue, New York, NY 10016

Library of Congress Cataloging-in-Publication Data
Clayton, Blake C. (Blake Carman), 1982–
Market madness : a century of oil panics, crises, and crashes / Blake C. Clayton.
    p.   cm.
Includes bibliographical references and index.
ISBN 978–0–19–999005–4 (hardback)
1. Petroleum industry and trade—United States—History. 2. Petroleum reserves—
United States—History. 3. Energy consumption—United States—History.
4. Energy policy—United States—History. I. Title.
HD9560.5.C5675 2015
338.2'72820973—dc23
                                            2014024750

1 3 5 7 9 8 6 4 2

Printed in the United States of America on acid-free paper

CONTENTS

PREFACE

This book is about what you might call the flip side of irrational exuberance—"irrational anxiety" might be the right term, or perhaps "irrational pessimism." Like "irrational exuberance," the phrase that Federal Reserve Chairman Alan Greenspan famously employed in 1996 to describe an overheated American stock market, irrational anxiety also refers to a certain kind of popular fervor that can accompany a bull market—but for oil, not stocks.

Unlike the stock market, where rising prices are associated with rosy visions of economic growth, boom times in the oil market often feature a growing chorus of frightening predictions that the end of oil is nigh. The mood is apocalyptic, not optimistic. The world is running out of oil, some alarmists cry. The planet can pump no more; oil production is doomed by geology to shrink from here on out, leading prices to rise as far as the eye can see as the commodity becomes more and more expensive to extract. In the oil market, unlike most other markets, this unflinching pessimism, coupled with a forecast that demand for the stuff will grow quickly, is the ultimate bull market argument. Anxiety about the future of oil becomes the ultimate reason to buy, and thus the ultimate boom-time story. But it has always proven wrong. True, oil prices can rise for stretches of time. Yet oil production globally has never hit the wall the Cassandras swear is right around the corner, nor has a bull market for oil ever not given way to bear.

This book tells the story of episodes of widespread fear in the United States about an imminent, irreversible collapse in oil production, which some prominent voices feared would cause oil prices to rise for a prolonged period and perhaps indefinitely, or at least until oil demand declined. These fears were not always irrational in a pure, academic sense. They were often the product of limited information about the world's oil reserves, which ultimately proved far too conservative, as well as a lack of appreciation for the ability of market forces and technological advancement to change the dynamics of reserve calculations.

That said, history points to a persistent bias among analysts to extrapolate to-day's price trajectory too far into the future, interpret rising prices as evidence of oil running out for good, and discount too highly the likelihood of high prices to stimulate additional oil production over time.

The purported "end of oil" has been a hallmark of popular debate about energy over the last decade. In 2008, as oil prices raced past $100 per barrel, anyone who turned on CNBC, picked up a newspaper, or skimmed a blog about the future of energy would have almost certainly been introduced to the idea that the world's oil supplies were simply giving out, coupled perhaps with the recommendation to buy oil until that far-distant day when, if we were lucky, mankind might develop enough alternative energy sources to spare prices moving higher and higher each year. Such predictions, which had been making headlines over the half decade prior, looked downright prophetic when oil eventually hit a once unthinkable peak of nearly $150 per barrel in the summer of 2008. Public opinion polls showed that most Americans believed that the world was indeed running out of oil. Ever-rising prices for gasoline on street corners around the country seemed to corroborate that view, as did countless documentaries, books, articles, and websites with terrifying titles like *Out of Gas: The End of the Age of Oil* (2004), *The Final Energy Crisis* (2005), *A Thousand Barrels a Second: The Coming Oil Break Point* (2006), *PetroApocolypse Now?* (2008), and the most aggressive forecast of all, *$20 Per Gallon: How the Rising Cost of Gasoline will Radically Change our Lives* (2009)—the *Dow 36,000* of the commodity boom, though its pessimistic inversion.

According to this alarming chorus, the world was running out of oil to pro-duce, leaving prices headed irrevocably in one direction—higher—the victim of unforgiving geology and surging global appetite for energy. It would be the boom to end all booms (and perhaps even the end of civilization itself, accord-ing to some of the bleaker prognosticators of the time).

It did not turn out that way. Oil production did not plateau, let alone plum-met. Instead, production has continued to reach new highs nearly every year since 2008. Yes, oil prices did rise for much of the 2000s, hitting sweltering new highs. But the collapse they suffered in the second half of 2008 was equally dramatic. In a stunning reversal, oil prices fell through the floor, nearly touch-ing $30 per barrel. Anyone betting on riding a perma-bull market off into the sunset of an early retirement did not end up even leaving the ranch. Just two years later, though, world oil prices recovered, rebounding back into triple-digit territory. Prices in 2012 were the highest they have been, on average, since the foundations of the modern oil industry roughly 150 years ago, in both nominal and inflation-adjusted terms.

Such lofty prices are in no small part the result of supply constraints. There are limits to the Earth's resources. Yet today's prices are driven as much by non-geological, above-ground factors (the limited willingness of a handful of countries rich in oil reserves, whether in OPEC or otherwise, to invest in new production capacity even as they sit on the lion's share of the world's best supplies, not to mention instability in vital oil-producing countries like Libya, Iran, and Iraq) as they are by geology. And there is reason to doubt that the record-high prices of the last half-decade will last indefinitely. Crude oil production worldwide is surging higher like never before, undergirded by a tidal wave of oil from North American fields that has taken even the most experienced oil experts by surprise. Oil remains expensive, yes, but the long-prophesied ceiling in global production looks farther off than ever, and the supposedly interminable upswing in prices of the last decade has proven, once again, not so interminable. And so—at least as of the time this book went to press—Western civilization remains intact.

Robert Shiller, the Nobel laureate Yale economist whose book *Irrational Exuberance* correctly called the dot-com bubble of the late 1990s (and the housing market in 2005, in a second edition), noted that every major boom in the stock market over the last century had been accompanied by what he calls "new era economic thinking." New era thinking is the "popular perception that the future is brighter or less uncertain than it was in the past." It takes the form of a widespread belief that the economy has or is entering a new epoch, thanks to some fundamental change that will lead asset prices to continue rising steeply for the foreseeable future, and perhaps perpetually. New era theories are often sparked by significant technological breakthroughs, like the widespread introduction of the television or the Internet. But people get carried away.

When new era economic thinking takes hold, investors come to believe that the progress justifies terrific gains in the market with no end in sight. Experts weave after-the-fact justifications for rising prices, and rising prices themselves appear to the public to be compelling evidence that such theories are correct. People caught up in new era theorizing fail to identify the basic similarities between the latest stories and similar ones that have taken hold in the past. It looks to them as though the normal laws of financial gravity no longer apply, and this time really *is* different. Inevitably, however, the boom ends, and prices cease rising, or even collapse. When they do, the new era theories that made for such mesmerizing cocktail party talk a few months earlier look rather silly in hindsight.

New era economic thinking is not limited to the stock and housing markets, however. It is a phenomenon that has come and gone over the boom-and-bust history of the American oil market as well. Swelling demand for oil and rising prices set into motion a wave of predictions about a new era in which oil will be in short supply. Some experts claim the imminent shortage will be permanent, or close to it. They question whether today's rates of oil production can ever be increased, or if the world's reserves have passed the tipping point toward exhaustion. Some of these analysts go so far as to prophesy that oil is running out once and for all. Such predictions are often coupled with forecasts that see oil prices rising far out into the future—even perpetually, unless and until consumption declines—convinced that lower prices are a thing of the past.

Prominent officials and energy experts are not immune to popularizing such gloomy new era views, which make for excellent copy for a news media hungry for tantalizing headlines. Oil industry executives often speak out against these views, dismissing the shortage speculation as unwarranted, but public skepticism of the industry means its rebuttals tend to fall on deaf ears. New era narratives have tended to thrive during periods when many Americans were concerned about the sustainability of oil as an energy source. Widespread frustration with an oil-based economy, which typically occurs when gasoline prices are high and rising, seems to accompany speculation in the news media that oil supplies will soon reach their limits, to be replaced be other, cleaner sources.

New era thinking is not limited to predictions of endless shortage, however. It can also manifest itself in the rosy visions of those who see unending abundance. When oil production is booming, inventories are rising, and prices are falling, the temptation to expect these conditions to persist far out into the future can prove almost overwhelming to market analysts. A few years of low oil prices can be enough to stir ill-judged talk of a new world of cheap oil—low prices as far as the eye can see. Yet history has shown such unflinching optimism to be just as dangerous as its more pessimistic counterpart.

Perhaps at no time was such new era optimism more on display than in the wake of the Asian financial crisis of the late 1990s. With some grades of crude oil selling for less than $10 per barrel by the end of 1998, commentators had become enamored of the futuristic array of information technology and other market-altering developments that would lead to a new era of low oil prices. The "industry has changed fundamentally since previous shocks that seemed to portend ever higher prices," two analysts told *Time* in April 1998. "If you are having haunting visions of long lines and $2.50-per-gal. gasoline, relax."[1] Yet just four years later, oil prices would begin an upward surge as fierce and consequential as any since the foundations of the modern oil industry. By the summer of 2005, U.S. gasoline prices had reached $2.50 per gallon, despite earlier predictions to the contrary. Prices have rarely fallen below that threshold since that time.

Yet predications of dire shortage and rising prices have received a degree of public attention over the course of American history that their more optimistic (though no less foolhardy) counterparts could only dream of, perhaps because they make for better newspaper copy. This book tells the story of four waves of new era anxiety about a looming end to oil reserves, and with it permanently rising oil prices, which have swept the United States over the last hundred years. The first episode lasted from roughly 1909 until 1927, set in motion by bleak assessments of domestic oil resources released by the United States Geological Survey, coupled with a take-off in American fuel consumption. U.S. officials and geological experts at the highest levels held little back in warning the public that the country's reserves would soon run dry. The second wave of new era theorizing about a coming collapse in oil production and permanently higher oil prices occurred between 1940 and 1949. The Allies' seemingly unquenchable need for oil in World War II appeared to many U.S. officials to herald a new era of permanent oil shortage, causing Washington to publicly fret that domestic oil reserves may run dry in a few years' time. More than two decades later, in the 1970s, similarly intense fears returned, catalyzed by the two oil crises of that decade, the first in 1973–74 and later in 1979–80. This third episode was characterized by worry among the public and in the Oval Office about never-ending oil price increases and a catastrophic collapse in oil production. It was front-page news and a powerful force in the cultural pessimism of that time in American history.

The latest new era narrative was reborn at the outset of the twenty-first century, now in its fourth iteration, as outages in major OPEC suppliers and astonishing increases in demand from developing countries like China caused the market to tighten sharply. With rising prices came a growing unease among many observers that a permanent decline in global oil production, and with it an unending rise in oil prices, may have arrived (or would very soon). A school of thought known as "peak oil" helped popularize this view. Peak oil advocates argued passionately that global oil production had bumped up against an unshatterable ceiling, leaving prices to rise more or less indefinitely until consumption declined. The fate of oil-driven industrial society itself was hanging in the balance, peak oilers argued, as the oil world was reaching its long-prophesied limits to growth. But when oil prices collapsed dramatically in 2008, the popularity of the peak oil view started to wane. By 2013, as global oil production globally continued to climb, driven by a remarkable renaissance in American oil production that continues to gain steam, defenders of the old idea of "peak oil" were almost nowhere to be found.

No "new era of shortage" scare has lasted. Higher prices expand the amount of oil below ground that can be profitably drilled, thereby spurring additional production. They also provide the carrot that risk takers need explore for oil in new ground and in new ways. Improvements in technology help companies locate new resources and extract oil more cheaply. High prices also mute oil demand, causing people to consume less of it and make use of substitute energy sources where possible. They find ways to use oil more efficiently. When oil production hits new highs, and prices inevitably fall, the chorus of voices that once trumpeted the new era of higher oil prices and long-term shortage fade back into the woodwork, just as apostles of the latest stock market craze do when things go south. Just as rising oil prices popularize the view that the world is facing a critical long-term oil shortage, or even running out of the stuff entirely, falling prices help drive the shortage narrative out of mainstream conversation. The same newspapers that once heralded a new era of long-term shortage go silent on the subject—or simply refocus their coverage on the astounding new oil production boom underway!

Popular discussion of the idea that the oil world is running out of oil is closely correlated with crude oil prices. When prices climb for prolonged periods of time, talk of the "end of oil" picks up. When prices eventually fall, the idea loses currency. This striking correlation is more than just anecdotal. One way to get a back-of-the-envelope sense of the quantitative link is via a service that Google Books offers called Ngram Viewer, which displays a graph of how often a given phrase appears in the vast corpus of books, government records and bulletins, and other reports that the Silicon Valley giant has digitized.[2] In figure I.1, the heavy black line shows the inflation-adjusted world benchmark

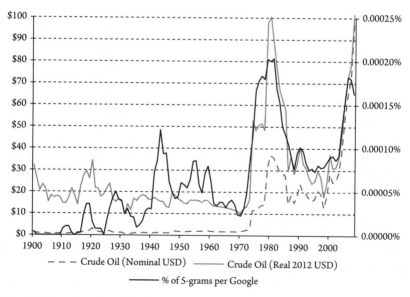

Figure I.1 Benchmark crude oil prices vs. mentions of running out of oil/gasoline, 1900–2008.

price of crude oil between 1900 and 2008, according to the energy company BP.[3] The dotted grey line shows the frequency of the phrase "running out of oil" or "running out of gasoline" among the entire body of five-word phrases in the English-language Google Book collection.[4]

The close relationship between the number of times "running out of oil" and "running out of gasoline" appear in print, and the price of oil, is striking, particularly since the 1970s. Oil prices and public attention to the idea that oil is running dry have been highly correlated for much of the last century, particularly the last fifty years. Three of the four shortage scares chronicled in this book—in World War I and the Roaring Twenties, World War II and its aftermath, the oil crises of the 1970s, and the peak oil craze of the 2000s—all coincided with climbing oil prices. Of these, only the World War II era did not come about in tandem with a surge in prices, likely because prices were strictly controlled by the U.S. government. Shortage fever was apparently in the air, though, as newspapers and periodicals of the era make plain. This simple exercise using Google Ngrams is more suggestive than definitive. But it underlines the fact that the price of oil, and stories about running out of oil, go hand in hand.

Looking at a second data source corroborates this pattern. Figure I.2 plots benchmark crude oil prices against mentions of "run out of oil" or "running out of oil." Rather than drawing on written sources in the Google Books

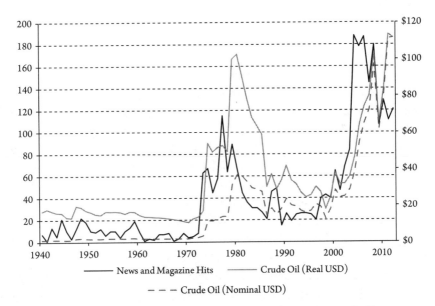

Figure I.2 Benchmark crude oil prices vs. mentions of "running out of oil" in newspapers and magazines, 1940–2012.

collection, data for this figure are from major world newspapers, magazines, and trade journals in the English language. These sources span the *New York Times* to *Motor* magazine, *Oil and Gas Journal* to the *Times of New Delhi*, and the *Economist* to *Time*. Because the number of sources was thin in the first half of the century, the figure shows from 1940 onward, ending in 2012. Perhaps even better than the Google Books library, newspapers and magazines give a good approximation of what normal, educated people in the English-speaking world were thinking and talking about at the time. The pattern shown in the graph bears a close resemblance to the one in figure I.1. In times of rising oil prices, conversation about the "end of oil" was rampant. When oil prices were falling, the idea fell off the radar. Apparently oil was no longer "running out"— despite the fact that more and more had been pumped and consumed, cumulatively speaking.[5]

Oil shortage scares thrive by extrapolating recent stagnations in oil production and rising oil prices far out into the future. Although such simplistic forecasts never come true, they can enthrall a wide audience for a time. Once inflation is factored in, prices have ebbed and flowed wildly since Colonel Drake drilled the country's first commercial well in 1859. Yes, there have been long stretches of time, even decades, in which prices have risen, but pessimists' visions of ever-rising prices at the so-called end of the oil age have never materialized. Dire claims of perpetual shortage have always proven wrongheaded. Likewise prices have never climbed straight upward, despite the fact that oil is being "used up" at an accelerating rate.

All of this is not to say that the oil market has not undergone significant structural shifts over the last century, which have had significant implications for prices and production patterns. Over the course of the modern oil industry, there *have* indeed been moments of fundamental change in the structure of the market, which have ushered in a new era of altered price dynamics. The implementation of mandatory production rationing by the Texas Railroad Commission in the 1930s and later the rise of OPEC as a restraint on oil exploitation in large swaths of the world in the 1960s and 1970s are both examples. Both events heralded new epochs in the oil market, as far as prices were concerned. The long history of oil supply and prices are different from those of many other assets in that respect. The U.S. stock market, for example, has arguably not undergone an analogous structural shift over the last century. The global oil market, however, has undergone major shifts in how oil itself is priced—today's complex web of financial contracts traded in the open market whose prices are benchmarked against various physical oil streams being only the latest iteration.[6] The ebb and flow of political power among major producing and consuming countries, the organization of and balance of power in the world oil industry, and geological inflection points have all resulted in seismic shifts in the very structure of the world oil market and pricing system over the last century and a half. New eras in the oil market, in this sense, have come and gone since the foundations of the market in the mid-nineteenth century.

Yet new era theorists, time and again, have made two critical errors in interpreting these historical punctuation marks. The first mistake is one of overidentification—in other words, they have claimed to see a structural shift in the market where there was none. Peak oil theorists of the mid-2000s are guilty of the first error. The paradigm is built upon the notion the world would soon hit peak production due to the fact that "approximately half the total oil available has been pumped," in the words of its founders.[7] This case for peak oil implies that a certain irreversible geological shift has occurred: mankind has crossed the Rubicon into a last remaining 49 percent of its oil resources, at which point a chronically worsening depletion sets in.[8] Yet no such Rubicon existed; no "midpoint of depletion" was ever crossed, though new era thinkers made their case on exactly such a threshold having been irrevocably stepped over. The world continues to produce and consume more oil than ever before.

The second new era error is identifying a legitimate structural shift but wrongheadedly arguing that the change implies that oil prices would rise far out into the distant future, or perhaps perpetually. The new-era-of-oil fervor of the 1970s provides a good example. When OPEC assumed the reins of the market in the early 1970s, it marked a structural shift in world oil supply patterns, as well as for oil prices. The price of a barrel of oil since 1973 has, on average, been higher and more volatile than it was for the preceding four decades.[9] Some observers, such as President Jimmy Carter, saw—correctly—that the world oil trade was undergoing an important change. But what the change would mean for the market over the very long term was frequently exaggerated and overinterpreted by analysts, in a fit of new era frenzy. In Carter's now-famous "unpleasant talk" with the American people in 1977, he expressed the belief that the oil crisis would not be over "in the next few years, and it is likely to get progressively worse *throughout the rest of this century*" (emphasis mine). It was mathematically impossible for oil production to climb any higher. "Just to stay even," the president reasoned, "we need the production of a new Texas every year, an Alaskan North Slope every nine months, or a new Saudi Arabia every three years. Obviously, this cannot continue."

Yet what happened in the years following? World oil production roared higher. Nominal oil prices, which had risen from $1.80 per barrel to $36.83 between 1970 and 1980, took a nosedive. Prices fell every year for the next six, collapsing to $14.43 per barrel in 1986. And they did not stop there. Real oil prices went on to decline throughout the 1990s, from $40.83 per barrel (in 2011 dollars) in 1990 to just $17.55 in 1998. OPEC was still in business a decade after Carter's dreary prophecy—barely—but its continued existence hardly meant a dire shortage of oil. Nor did it mean a straight-shot upward for oil prices, as new era thinkers foresaw in the dregs of the late 1970s. The end of oil prophesied once again proved elusive.

Is This Time Really Different?

Built around four historical case studies, this book explores the conditions under which popular predictions of an imminent, long-lasting shortage of oil arise, gain popularity, and eventually wane. The book documents a recurring bias toward pessimism about future supplies and prices, which have been evident over the last century of oil. These widespread yet error-prone beliefs in an impending shortage—capable of causing prices to rise indefinitely—have appeared in discussions about U.S. energy policy and markets over history. By telling the story of these short-sighted scares, this book calls into question some of the dire predictions about future energy supplies that linger today, by asking—"Is this time really different?" Can it be that the prophets of permanently declining supplies and ever-higher prices who made the headlines over much of the 2000s will be proven wrong yet again?

The message of the book is similar to Carmen Reinhart and Kenneth Rogoff's *This Time Is Different*. In that book, the two Harvard economists distill eight centuries of data on financial crises down to one basic insight: "Our basic message is simple: We have been here before. No matter how different the latest financial frenzy or crisis always appears, there are usually remarkable similarities with past experience. . . . Recognizing these analogies and precedents is an essential step toward improving the global financial system."[10] The same wisdom applies to the realm of energy, especially when it comes to oil. No two chapters in history are identical, and yet distinguishing the patterns of the past is a crucial starting point for distinguishing false alarms from real historical inflection points.

This book is not an attempt to predict the future. There is no guarantee that oil production globally will continue to rise indefinitely in the years ahead. A more likely scenario is that a prolonged period of high prices will eventually push people toward other energy sources, which will lead to less and less economic appeal in finding new and better ways to extract crude oil. Oil will be outmoded long before the world's oil wells run dry. If the long history of oil teaches anything, is that the oil market is marked by sharp disjunctures and discontinuities, which have confounded most attempts to peer beyond a few years into the future. Yet, as foolish as it would be to extrapolate past trends into the future, it would be equally unwise to disregard the analogues and precedents of history to understand the forces that shape today's market, and will likely characterize tomorrow's. Human forgetfulness is the ultimate source of financial folly.

Unlike Shiller's works on irrational exuberance in the U.S. equity and housing markets, this book does not argue that oil prices are experiencing a bubble. Although exceptionally high by historical standards, oil prices (as of the time this book was written) appear justified according to the fundamentals of supply and demand. Most econometric research suggests that prices were equally plausibly justified by market fundamentals between 2003 and 2007, as oil prices surged. If there were a time that the world "bubble" would have best applied to the oil market, it would have been for a stretch of time in 2008, as discussed in the Introduction.[11] In any case, a lack of comprehensive, high-quality data about the flow of oil in world markets makes it more difficult than most people realize to determine the degree to which prices at a given point in time are deviating from reasonable opinions of fair market value. That is particularly true in times of exceptional economic uncertainty, such as the latter half of 2007 and 2008.

Too often, popular discussion about the future of finite resources like oil is limited to tired debates of scarcity versus abundance. Although some historians have noted that misguided predictions of an impending shortage dot the history of oil, they often cite them only in an attempt to cast doubt on similar forecasts today. The fact that such predictions were wrong in the past does not mean that similar ones today are necessarily wrong, too. Yet it does justify reasonable skepticism about why today's prophecies should be more successful than those failed ones of the past, and why methodologies that have proven inept in an earlier era will generate more accurate results this time around.

As of the time of the writing of this book, fundamental supply-demand forces remain locked in a pitched battle, though crude oil's rapid slide in the autumn of 2014 has pulled the market well below the triple-digit-levels to which the world had become accustomed. Prices still remain high by historical standards, however. Chinese-led demand growth in the emerging world continues to bump up against global production capacity. Many of the world's major reserve holders set fiscal and regulatory terms that hinder investment or are hampered by geopolitical disruptions, causing production to fall far short of what it could be and keeping oil prices aloft. Yet there is no quick solution for this deeply entrenched social—not geological—impediment to the market. Despite the Great Recession, demand growth for oil worldwide remains fairly strong. Moreover, the cost of producing additional barrels of oil outside of OPEC is currently not as cheap as it has been in epochs past, when the United States could produce more than enough oil for itself at a fraction of today's prices.

Yet a transformation is underway in the world oil market whose ultimate unfolding may end up causing today's scaremongers to look as foolish as they have in generations gone by. The same sweeping pattern of history—of large price increases leading to widespread new era shortage fears that eventually dissipate when oil production eventually rises and prices moderate—appears to be playing itself out again since prices exploded and collapsed in 2008–9. In the years leading up to that epic $147-per-barrel spike, the view that the end of U.S. oil production was in sight was commonplace. But the opposite has happened: the country's output has risen 70 percent since that time, and the International Energy Agency (IEA), the West's energy watchdog, estimates it will increase a further 30 percent by the end of the decade. The seeds of a secular shift in world oil production, pointing toward supply gains previously uncontemplated, have been sown.

What is behind the change in outlook? High prices have spurred breakthrough gains in technology, like the combination of hydraulic fracturing with horizontal drilling, and made possible plays that were previously uneconomic feasible. These advances have sparked a revolution in U.S. natural gas production and made it the largest in the world, surpassing Russia. The techniques have also been successfully deployed to produce so-called tight oil, extracting liquids from dense rock. Replicating these gains in other parts of the world will happen only slowly. Yet the tremendous economic incentive that countries outside North America have to follow suit will likely prove irresistible over time, and the flood of new oil will continue, helping to make up for declining output in some areas. Meanwhile, oil consumption in the United States and Europe has stagnated or even declined since the Great Recession, owing to prohibitively high prices at the pump, demographic shifts, and stricter fuel economy standards for cars and trucks.[12]

The deeply cyclical nature of the world oil market, like other commodity markets, is on display as vividly right now as at any time in its century-and-a-half-long history. Time will tell whether oil prices will continue to oscillate around the $85-per-barrel levels of the autumn of 2014, whether they will grind higher on continued geopolitical deterioration in the Middle East, whether the rising tide of U.S. and Canadian oil and gas production will continue to mark the first chapter in a new bear market—or whether the forces of history will conspire in a different direction altogether.

ACKNOWLEDGMENTS

I could not have asked for a better professional setting for writing this book, which began as my dissertation at Oxford University and developed during my time at the Council on Foreign Relations (CFR), than what Richard N. Haass and James M. Lindsay provided me at CFR. It afforded me the resources and time I needed to write and re-write. The combination of autonomy and guidance they provided me, as a first-time author, was invaluable. Special thanks also go to Michael Levi, whose broad energy expertise and research acumen refined my thinking.

Jason Bordoff, Charles Ebinger, David Goldwyn, Faisel Khan, Michael Levi, James Lindsay, Bob McNally, Kenneth Medlock, Edward Morse, Amy Myers Jaffe, Adam Sieminski, and Daniel Yergin all read the manuscript and generously shared their insights. I am also grateful to other colleagues at CFR whose advice I leaned on: Amy Baker, Max Boot, Patricia Dorff, Elizabeth Economy, Robert Kahn, Sebastian Mallaby, Shannon O'Neil, Adam Segal, and Benn Steil. Alexandra Mahler-Haug provided exceptional research assistance.

Citigroup was gracious and accommodating with regard to the project since I joined the firm in 2013. I owe a special thanks to Jonathan Rosenzweig, Jon Rogers, and Michael Artura for their support. I am also grateful to Faisel Khan for sharing his encouragement and feedback. His wealth of industry knowledge and capital markets expertise were pivotal. Edward Morse, who has been a generous mentor and colleague, also provided critical insights. His perspective on the industry is one of a kind.

The book also benefited from the guidance I received from my former colleagues at Oxford, who helped to shape the manuscript in its earliest stages. David Barron, my doctoral supervisor, gave me thoughtful feedback and perspective on the research and writing process. The Centre for Corporate Reputation provided generous research support for my doctoral work, for which I owe a special thanks to Rupert Younger, its founder and director. Dana

Brown, Ray Loveridge, Paul Stevens, Eric Thun, and Marc Ventresca helped nudge my first few drafts in the right direction. Thanks also go to Bassam Fattouh and the Oxford Institute for Energy Studies.

Lisa Adams, my agent, was a trusted counselor from start to finish. I also owe thanks to Scott Parris, my editor at Oxford University Press, who championed the book from day one. Cathryn Vaulman was patient and skillful in overseeing the publication process.

Two people deserve special mention. The first, Matthew Simmons, introduced me to the oil and gas industry when I worked with him on a private equity venture in Maine and Houston in 2010. Sadly, Matt passed away that summer, so our time together was cut short. But his passion for the oil market and his love of a good debate were instrumental in shaping this book. Most of all I owe thanks to my wife, Amy, for her unwavering support during the ups and downs of getting my doctorate, moving across the country (and the Atlantic), transitioning to CFR and Citigroup, and writing this book. She is an extraordinary person and friend.

This book was made possible by the generous support of the Alfred P. Sloan Foundation as part of the Council on Foreign Relations' Project on Energy and National Security. I alone am responsible for the book's contents.

1

Introduction

"But what if stories themselves move markets? What if these stories of over-explanation have real effects? What if they themselves are a real part of how the economy functions? . . . The stories no longer merely explain the facts; they are the facts."

Akerlof and Shiller, *Animal Spirits*, p. 54

In 1863, the Pennsylvania oil boom was at full tilt. The speculative excitement in the region was palpable, bringing would-be titans of the nascent industry to the fields in droves. Local newspapers hailed it as the American El Dorado. So much oil flowing was a nuisance, as well as a blessing, contaminating wells and leaving a thin, bluish coating on the creeks. But to fortune-seekers, it was a scene of unmitigated plenty. This new rock oil was cheaper, more plentiful, and cleaner burning that any of the alternatives, and the perfect source of illumination. One of the world's most powerful and lucrative industries was just being born.[1]

An enterprising Scottish immigrant, Andrew Carnegie, wanted in. Carnegie would eventually become one of the richest Americans of all time, with a net worth in today's money just shy of $300 billion by some estimates. But for now, his standing in society was more humble. As the division superintendent of the Pennsylvania Railroad, Carnegie had a bit of extra cash to invest and access to several prominent Philadelphia businessmen with deep pockets. William Coleman, a business associate of Carnegie, had purchased a tract of oil land on Pennsylvania's Oil Creek in 1859, not far from where the first American oil well had been drilled that same year. Coleman formed the Columbia Oil Company to drill for oil on the land and invited Carnegie to get in on the ground floor.[2]

Wanting to learn more about the venture before putting any of his own money into it, Carnegie said he was interested, but asked to see Oil Creek for himself before he bought into the scheme. What he saw there astounded him. The local scene was one of almost total chaos. Makeshift oil derricks covered the land, only feet from each other. Prospectors huddled in tiny wooden shanties—at least those who were lucky enough to have a roof over their heads at all. Much of the drilling equipment and rough-hewn dwellings were a charred mess due to rampant oil fires. Notwithstanding the devastation, the allure of wealth was palpable. Apparently it was more than enough to survive on. "What surprised me," Carnegie later reflected, "was the good humor that prevailed everywhere. It was a vast picnic, full of amusing incidents. Everybody was in high glee; fortunes were supposedly within reach; everything was booming."[3]

Carnegie decided to invest. His new partner was delighted. But they were hardly going to pursue the same drill-baby-drill strategy that everyone else was. Unlike all of the men drilling at a breakneck pace, they would take the opposite tack. They could tell a speculative frenzy when they saw one, and neither wanted any part of it. Instead, they decided to make the contrarian play. And that is when things got interesting.

Ever the realists, Carnegie and Coleman knew what was coming: The day when all the world's oil would be gone forever. Why pump our field dry right now, they reasoned, when we could make a killing by selling the stuff after everyone else's oil has run out? That would be the winning strategy: Await the "not too distant day," in their words, until "oil supply would cease." They could simply pump a little now, and then hoard the oil until the worldwide oil shortage had struck. It probably would not be long before they were the only ones with any left. After all, these fields could not hold up forever.

All they needed was a place to store their oil. They decided to dig an enormous hole, capable of holding 100,000 barrels of oil, where they could hole it away. They did not waste any time. Before long, and thanks to a lot of digging by some hired help, their "lake of oil" was ready. They immediately filled it with the production from their newly bought oil wells, all 100,000 barrels worth. It was there, in the lake of crude oil, that the business partners would stash the oil until the supply crunch they awaited had sent prices soaring.

It seemed to be a brilliant plan. They would have the market cornered and become the kings of oil. Coleman reckoned that "when the supply stopped, oil would bring $10 a barrel and therefore we would have a million dollars' worth in the lake." One million dollars—it was almost too easy. Now all they had to do was wait for everyone else's wells to run out. They apparently saw little danger of oil being discovered elsewhere. So they waited. And waited. But the long-awaited shortage never came. The only thing that did arrive was evaporation, which kept skimming more and more oil off the top of the lake. "After losing many thousands of barrels waiting for the expected day . . . we abandoned the reserve." They emptied the lake onto the market and got out of the hoarding business.

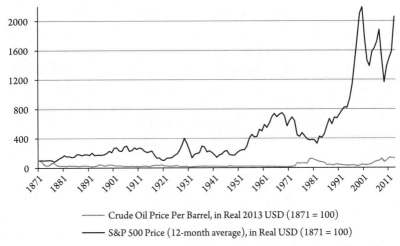

———— Crude Oil Price Per Barrel, in Real 2013 USD (1871 = 100)

———— S&P 500 Price (12-month average), in Real USD (1871 = 100)

Figure 1.1. World benchmark crude oil prices vs. the S&P 500, adjusted for inflation and set equal to 100 in 1871 (1871–2013). Source: *BP Statistical Review of World Energy 2013*; Shiller, historical S&P 500 prices and U.S. inflation data, which can be accessed at http:// www.econ.yale.edu/~shiller/data.htm.

Writing his memoirs in 1920—a full fifty-seven years later—Carnegie noted wryly that the shortage he and Coleman were awaiting still had "not arrived." Upon reflection, he attributed his foiled plan to a vital miscalculation: "We did not think then of Nature's storehouse below which still keeps on yielding many thousands of barrels per day without apparent exhaustion."[4]

Looking back at the scheme from the vantage point of today's triple-digit oil prices, it is easy to think that Carnegie's only mistake was selling his lake of oil too soon. Had he only (or more likely, his great-grandchildren) held out for another hundred years or so, some might think, his investment would have finally paid off hugely. In time, he would have been vindicated. Not so.

Even if Carnegie and Coleman could have cheated death and sold their lake of oil in 2014, their investment would still have been terrible, with no realistic odds of breaking even. Oil is not any more valuable today than it was during the bull markets of their time, once inflation is factored in. When Carnegie was drilling his lake of oil, likely around 1864, the average real price of a typical barrel of U.S. crude oil was $119 per barrel. In 2013, by comparison, a barrel of benchmark West Texas Intermediate crude averaged $98 and Brent crude sold for $109. The market was extremely volatile in Carnegie's era, so it is possible that oil prices were lower than that (they averaged $59 in 1863 in today's terms). Regardless, once the costs of storing the commodity is factored into the equation, he would have had no real chance of breaking even, let alone making a fortune. If long-term gains were what they were after, Carnegie would have fared much, much better by putting his money to work somewhere else, like the stock market. As figure 1.1 shows, a basket of stocks comparable to the S&P 500 dating back to 1871 is worth 21 times what it was back then. An initial investment of $100 would have grown to about $2,056 after inflation. Had that same initial amount of money been invested in crude oil, on the other hand, it would not be worth more than $130 today. Not exactly a rich reward for a 142-year wait. The lesson is simple but counterintuitive: A barrel of oil is worth about as much now as it was in Carnegie's era once inflation is factored in, despite the vast quantities of the stuff that have been pulled out of the ground since that time.[5]

Awaiting the End of Oil

It is easy to scoff at Andrew Carnegie's failed scheme to corner the world oil market by digging a pit in Pennsylvania, but the future titan of American industry was no fool. In fact, that is what makes the story remarkable: Here is one of the modern world's great industrial minds—not exactly your average investor—with a carefully laid plan, a seemingly reasonable investment strategy, and the means to carry it out. And yet, in hindsight, his miscalculation was laughably wrong. One hundred and fifty years later, the global oil production collapse that he and his colleagues were banking on still has not struck.

What makes Carnegie's youthful misjudgment so interesting is that he is hardly the only one over the course of American history who has confidently predicted that oil supplies would soon run out, causing prices to leap higher— quite the opposite, in fact. Pessimism about the future availability of oil is as old as the industry itself. Carnegie succumbed to it, but so have countless others, including those in several generations after him. The same prophecy of exhaustion that led him to dig the "lake of oil" is shared by many in the world today, who see humankind as doomed to suffer the negative consequences of geological stubbornness. Those who believe that the world is facing a permanent shortage of oil, or even an era of permanently higher prices, are only the latest voices in a timeless American refrain: "Oil is here today, but will be gone tomorrow," the song goes. It is a prediction with a long pedigree.

Fast forward to the summer of 2008. Oil prices were defying gravity. To American SUV owners, the market seemed like some sort of dystopian casino, where anyone who needed to buy gasoline stood in fear of what tomorrow's trading on the commodity exchanges would bring. It was almost so bad that you could not look away: $87 per barrel for Brent crude oil in January, $100 by February, $109 in March, then $115 the next month, followed by $129 in May. Those numbers may not sound high today (which is a troubling thought), but oil had never cost more than even $80 per barrel before 2006. And now prices were hitting $147 per barrel in July 2008.

But the ticker tape drama was not what was oddest about the market as prices ascended. It was what the experts were saying about *why* prices were rising. Pundits touching every corner of finance, energy, and politics—from hedge funds, investment banks, and the oil majors to OPEC ministers, the White House, and the Federal Reserve—were all being asked that question. They each had their own arguments about why prices were high and where they might be headed. Listening to them, it was clear that they frequently understood the most basic aspects of the oil market almost totally differently from each other. That was true whether the question was how much oil the world was producing, how much oil production capacity existed in the world, if oil prices would return to their previous lows, whether Wall Street was to blame for driving prices higher, and perhaps most fundamentally, whether publicly available data about the oil market could be trusted at all.

And then there was the most fundamental issue of all: Whether this was the end of the oil age—if the world was running out of cheap oil, once and for all, making $150-per-barrel oil the new normal.[6] On that point, two of the most widely cited oil experts, Daniel Yergin and Matthew Simmons, were a lesson in contrasts. Both were prominent within the oil industry and had deep experience in the market. Yergin, a Pulitzer Prize-winning historian, was chairman and founder of Cambridge Energy Research Associates, now IHS CERA, a leading energy consulting firm. Simmons was among the industry's most respected financiers, the co-founder and chairman emeritus of Simmons & Co. International, an oil and gas investment bank. Yet despite their common industry backgrounds, the chasm between where they saw the market going could not have been wider.

The two were on opposing sides of a very simple idea: Could the world continue producing more oil, or was society facing an inevitable decline in oil production—and with it the end of oil as a viable energy source? Yergin was optimistic that human ingenuity and economic incentive could fend off any chaotic shortfall of oil production in the future. Simmons, on the other hand, was less sanguine. Global oil production had reached its zenith (or would very soon). As a result, the price of oil was doomed to rise "so fast your head will spin." It was not a pretty picture. "I don't see why people are so worried about global warming destroying the planet," Simmons once said. "Peak oil will take care of that."[7]

Peak oil—the idea that the world's oil production was reaching "peak" levels and would not be able to rise any further—was an idea that began to bubble up in the late 1990s, at first from within the energy community. Colin

Campbell, a retired geologist whose career had included stints at several oil majors, described the problem of peak oil in a June 1996 essay:

> The peak of oil discovery was passed in the 1960s, and the world started using more than was found in new fields in 1981. The gap between discovery and production has widened since. Many countries, including some important producers, have already passed their peak, suggesting that the world peak of production is now imminent. . . . The world faces the dawn of the Second Half of the Age of Oil, when this critical commodity, which plays such a fundamental part in the modern economy, heads into decline due to natural depletion. . . . Petroleum Man will be virtually extinct this Century, and Homo sapiens faces a major challenge in adapting to his loss.[8]

In a 1998 essay for *Scientific American* co-authored with Jean Laherrère, Campbell predicted that oil production would likely hit an unsurpassable peak in 2004 of about 71 million barrels of oil a day (actual production was roughly one-quarter higher than that in 2012). Seven years earlier, he had published an obscure book arguing that the world's oil production would max out within ten years. The "epoch of declining production is about to begin," he warned, though he would later push back his date of its arrival.[9] The consequences of the upcoming energy crunch, as he foresaw them? Nothing short of a "permanent radical rise in oil prices."

Geologist Kenneth Deffeyes, a Princeton University professor emeritus and another early peak oil apostle, captured the spirit of the peak oil gospel in the opening paragraph of his 2001 book, which helped launch the movement:

> Global oil production will probably reach a peak sometime during this decade. After the peak, the world's production of crude oil will fall, never to rise again. The world will not run out of energy, but developing alternative energy sources on a large scale will take at least 10 years. The slowdown in oil production may already be beginning; the current price fluctuations for crude oil and natural gas may be the preamble to a major crisis.[10]

The most radical peak oil adherents argued that the world was on its way to running out of oil entirely, which would threaten modern civilization itself. Shortages of oil and stratospheric energy prices would do in the industrial world, they often argued. A more moderate strain of peak oil thinkers thought that global production would very soon or had already hit a peak, but they were not sure how bad things would be when that happened. To the peak oil crowd, Yergin was perhaps the most prominent example of a "cornucopian," a derisive label describing someone who thought that oil existed in enough abundance, markets tended to work well enough, and human ingenuity was great enough to prevent the doom and gloom that the peak oil adherents foresaw.[11] For peak oil's defenders, the dramatic rise in oil prices in the summer of 2008—with no sign of abating—was simply proof of just how wrongheaded their opponents' cornucopian views really were.

Many Americans had premonitions about the end of oil that were vaguely similar to Simmons's views. Hundreds of books, articles, studies, and other media put forward similar arguments during the mid-2000s. Experts and scholars wrote fretful works about the coming shortage, such as *The End of Oil: On the Edge of a Perilous New World* (2004), *Peak Everything: Waking Up to the Century of Declines* (2007), *The Party's Over: Oil, War, and the Fate of Industrial Societies* (2003), and *The Long Emergency: Surviving the Converging Catastrophes of the Twenty-First Century* (2005). A sense of foreboding about the country's oil economy was part of the zeitgeist of the George W. Bush presidency, no doubt driven largely by the oil association with terrorism and violence, whether in Iraq or Lower Manhattan, and rising prices at the pump. In one poll, conducted in April 2008, 76 percent of Americans said that "the world is running out of oil." A majority of respondents in fifteen of sixteen countries agreed.[12] America was coming to grips with the fact that it was "addicted to oil," as Thomas Friedman's 2006 Discovery Channel series termed it. And with the world's oil supplies unreliable at best, and already in terminal decline at worst, the addiction looked fatal.

Was this the beginning of the end for oil? It was not merely an academic question. Oil is central to the world economy. No other good on the planet is traded more, by market value. Transportation depends on it. Militaries require it. Economies rely on it. Yet, in the summer of 2008, the simple question of whether there was more oil or not in the ground, and if oil prices were destined to keep rising for the foreseeable future, remained an open question in newspapers and on television sets around the world.

In reality, 2008 was not the first time the world had supposedly "run out" of oil.[13] Prophecies of an imminent end to the oil age are as old as the oil industry itself. Like the boom-and-bust pattern that has defined oil production over the entire course of its history, so too have widespread, if misguided, predictions of ever-rising prices been a recurring feature of public discussion when it comes to the future of energy. These cries seem to get louder when demand outruns supply and prices rise sharply, only to be quickly forgotten when the market swings into surplus and prices inevitably retreat.[14] The same kinds of stories that helped to fuel the dot-com bubbles and housing crises of the last two decades also characterize the last hundred years of the oil market. Although these stories are not responsible for the zig and zag of oil prices, they can all too easily cause investors, politicians, and the public alike to grossly misunderstand what lies ahead for the world's energy supplies.

A host of sources—books, newspapers, magazines, the *Congressional Record*, government-commissioned studies, and later, documentaries, television shows, and blog posts—tell the story of these cyclical fears. Whispers of an impending oil shortage can be found in Pennsylvania in the 1870s and 1880s as the region's wells are tapped more and more aggressively. Carnegie's "lake of oil" came about in that context. Shortage fever took hold again in World War I and the booming 1920s, as the nation's love affair with the automobile began to take off, lifting consumption to once-unthinkable heights. The shortage narrative arose a third time in World War II and the 1950s. The thought that the country's oil reserves, recognized by the White House as a strategic commodity essential to modern warfare, would soon be gone struck fear into the heart of Washington's top brass. The narrative gripped the nation in the 1970s, as the oil crises of that decade appeared to many to herald the decline of the United States (and indeed of the West), which was running up against the limits to growth. Most recently during the mid-2000s, stalled output outside OPEC and a booming China catapulted oil prices higher accompanied by fervent claims that the earth's capacity to pump more oil was running out, if not the stuff altogether.

None of these eras was exactly alike, though they share certain striking similarities. History has not repeated itself, but it has rhymed. In each of these periods, expert predictions of an imminent, irreversible shortage of oil, broadcast widely by the media, have driven the scarcity narrative. Everyday Americans heard stories about a coming day when it would be more expensive to fill their gas tanks than ever before—if they were able to fill them at all. Deep cultural or political concerns often helped fuel the amplification of these worries. Undercurrents of fear about the future of the environment—for instance, about the need to conserve the country's natural resources or of irreversible harm being done to the landscape—often seemed to buttress the fears. So too did concerns that draining the country's oil fields would risk subjugation to a foreign military power, a fear often reinforced by the country's political and military elite. These narratives would hold sway in important circles for years or even a decade at a time, only to fade away, and then reappear once again a generation later. Each time, the end of the oil age appeared imminent. But in every case, a tidal wave of new oil eventually hit the market, lowering prices. The oil shortage story would disappear from the nation's media. It was boom-and-bust writ large.

This book tells the story of these waves of shortage fears, beginning at the outset of the twentieth century. It has two primary goals. The first is analytic. The study asks questions about the emergence, transmission, and decline of oil shortage narratives. Why do they occur in the first place? Do they appear randomly over the course of history, or are there reasons why they occur at some times and not others? Under what conditions do predictions of a long-term shortage of oil tend to crop up, gain popularity, and then fade away? The second goal of the book is prescriptive. Assuming there is a general pattern to the contours of these bouts of fear, what does it mean for consumers, investors, policymakers, and anyone wondering about the future of energy?

An Old Familiar Question

The question of whether finite resources will always prove sufficient to enable civilization to move forward is an age-old debate for economists.[15] Thomas Malthus, a British economist born in 1766, was obsessed with the notion that because there was only so much land that could be farmed, human population growth would one day hit an irreversible limit. Were people to multiply too quickly, they would lack the food to sustain themselves. Only an unpleasant brew of "vice and misery" could drag the total head count back to sustainability if his countrymen did not begin to practice "prudential restraint," which would limit population growth.[16] Ricardo, in his 1817 *Principles of Political Economy and Taxation*, responded to Malthus by arguing that there is enough agricultural land to go around—but its quality varies enormously from place to place.[17] While the most productive land tends to be worked first, inferior land (or mines, for that matter) can be made to yield comparable bounty if inputs are increased. That additional cost, though, would inevitably push the price of raw materials higher and higher, though it would also dampen consumption growth and population multiplication, softening humanity's fall.

But the generation following Malthus saw rays of hope where their predecessors did not. John Stuart Mill, writing in 1862, though accepting of the fact that resource constraints did pose problems for economic growth, thought that Malthus had missed the mark in two respects. For one thing, he argued, the ultimate quantity of arable land was much higher than his fellow economist admitted. It was a less pressing constraint than Malthus's writings led people to believe. What was more, he pointed out, technology can help. Mill was the first modern economist to see how better tools could help mankind deal with the problem of scarce resources. As technological progress marches on, he foresaw, so too can living standards improve even though some resources may become scarcer.[18]

William Stanley Jevons, another British economist, has the dubious distinction of trying to utilize the Malthusian framework to forecast the eventual fate of his island home. Writing in 1865, Jevons warned that the country's growing appetite for coal to power their booming manufacturing base, which he argued was unsustainable, would mean the end of Britain as they knew it. "The exhaustion of our mines," he prophesied, "will be marked . . . by a rising cost or value of coal; and when the price has risen to a certain amount comparatively to the price in other countries, our main branches of trade will be doomed." That would spell "the end of the present progressive condition of the kingdom . . . we cannot long progress as we are now doing." Alas, Jevons lamented, there was no source of energy found in Britain that could ever replace King Coal. Wind or tidal power, perhaps? Not reliable enough. Wood? Too unwieldy. Oil? This "supposed substitute" for coal was too "limited and uncertain" to be a good option.[19]

Jevons was right that his country's days as the world's largest coal producer were numbered. The reason was simple: It became cheaper to mine the stuff elsewhere, like the United States, then import it, rather than produce it at home. But he was dead wrong about its implications for the economy. Trade enabled the British to get what they needed when other countries, like the United States, began to outpace them. Lack of coal did not spell the end of Britain, nor did it mean the end of coal. Nearly 250 years since his prediction, proven reserves of coal have continued to multiply. England has long since moved on to other sources of energy for all but a fraction of its power—which is now cheaper, more abundant, and more reliable than Jevons ever contemplated.

By the turn of the twentieth century, however, economists largely eschewed the classical school's regard for scarcity as a fundamental constraint on growth. The rise of neoclassical economics, which were less prone to see absolute limits to natural resources as an irrevocable hardship for the upward march of the economy, constituted the economic orthodoxy for much of the twentieth century. It was a school of thought well suited to the times, given that the long-term trend in growth among the leading Western nations was promisingly upward. Even major economic collapses, like the Great Depression, did not seem to most economists as the product of natural resource constraints as much as they did speculative excess. Things healed and resumed growing. Although some notable cultural crusades, such as the American Conservation Movement, worried aloud about the fate of the country's natural environment if human treatment of it went unchecked, the mood in the academy was generally sanguine about what limited resources meant for the fate of society.

One dissenter from this chorus of optimists was Harold Hotelling, a Stanford statistician. His elegant model, on which much of the scholarship in the field of exhaustible resource economics is still based, predicts continually rising prices for finite resources, whether gold, oil, or otherwise.[20] The intuition is simple. People who own a resource that will one day run out will try to sell it at a time that maximizes what they can get for it. They know that if they sell it today, the chance to sell it tomorrow disappears. And they factor that into their asking price. If they stand to make more money by pumping the oil or mining the gold and then investing that money, rather than leaving the stuff in the ground, they will do that. The owner will calculate how much of the produce at any given time based on what the current marginal revenue it could provide, minus extraction costs, could yield as productive capital earning a market rate of interest. So wealth-maximizing oil producers, facing a fixed supply of oil in the ground, would extract the resource at a rate that ensured the market price for their oil rose with the Federal Reserve's discount rate over the decades, reflecting a so-called "scarcity rent."[21] Yet, however artful Hotelling's theory, it is at odds with observed behavior of prices over long periods of time.[22] Raw materials prices simply have not shown the steady upward climb that Hotelling would have predicted. Instead, the prices of most commodities have fallen, or at most been stagnant, over the centuries once inflation is factored in.

There are two ways to reconcile the Hotelling rule's prediction of rising real exhaustible resource prices with a reality that has proven far more hopeful. The first is that the market perceives the Earth's endowment of these vital resources as so vast, and their ultimate extinction so far in the future, that any actual scarcity rent the market attaches to them is a negligible component of their price. While the scarcity rent may be growing at the rate Hotelling would predict, extraction costs are overwhelmingly what determine real-world prices. Another way of reconciling the rule with actual price patterns is that technological gains have made it cheaper to produce exhaustible resources, eclipsing the net effect of any supposedly rising scarcity rent on prices.[23] It would not be impossible for the Hotelling formulation to show greater empirical validity at some point in the future for a given commodity, though, should market perception shift or a binding, measurable geological reserve-size limit come into view.

In the 1970s, the cultural tide would turn against neoclassical economists' rosier views of what human ingenuity and free markets could do to combat resource scarcity. A chorus of scholars—many of whose work was prompted by a combination of sharp increases in commodity prices during the decade, tectonic shifts in the geopolitical landscape (the rise of OPEC, for instance), and a resurgent American environmental movement—questioned whether neoclassical scholarship was too dismissive of the threat that resource exhaustion posed to global economic growth. The Club of Rome's *Limits to Growth* was perhaps the most culturally influential of these arguments.[24] A tiny pocket of economists tried to buttress these claims with solid analysis; more were probably sympathetic, but few made any attempt to reinforce such arguments with any rigor.[25] The economists who saw the commodity calamities of the 1970s as troubling omens of a larger quandary remained quite heterodox within the context of the larger economic community of the day and were generally dismissed as holding eccentric perspectives.

Mainstream economists firmly rejected the idea that scarcity limits would be a significant constraint on the heights the economy could reach. They found little persuasive in the Club of Rome's report.[26] One critique was typical: "With much fanfare and alum, Malthusian theories have recently been revived by a group of engineers and scientists." While such predictions "are impressive to laymen and scientists alike because they appear to be derived from sophisticated models," they are riddled with "subjective plausibility" and "represent no advance over earlier work."[27] In the neoclassical view, technological progress, coupled with the way in which the relative price changes between energy sources provide incentives for consumers to switch what they use and producers to swap what they supply, were sufficient to overcome perceived scarcity limits, keeping exhaustible resources from ever being totally consumed. Joseph Stiglitz, using a quantitative model, showed that at least three forces would always offset the supposed limitations imposed by natural resource constraints: technological improvements, substitution (of more scarce with less costly materials), and returns to scale (bigger producers can provide goods in short supply more cheaply). These forces make sustained growth in per capita consumption not only feasible, he argued, but the booming rates of resource extraction at the time were actually economically optimal.[28]

The future of the world's supply of crude oil was a flashpoint in the debate between neoclassical and neo-Malthusian scholars in the 1970s. Much of the neo-Malthusian thought about oil supply during this decade can be traced back to the work of M. King Hubbert, who predicted in 1956 that oil production in the United States would start to fall sometime between the mid-1960s and early 1970s.[29] Hubbert reasoned that oil supply would be exhausted over a period of time determined by the size of the resource base and how fast it was pumped. He assumed that the rise and fall of U.S. oil production would be neatly symmetrical, then arbitrarily drew a curve by hand to match the data to arrive at his famous 1956 prediction.[30] (He later fit a more precise bell-shaped curve to the data.) His message was clear: The country's wells could support more oil and gas production for a few more years, but then they would simply give out, never to rise again.

How did his famous forecasts fare? Hubbert's "high" estimate did correctly predict the period in which the country's production would peak (1970). But as time has gone on, his 1956 prediction has fallen increasingly off the mark. Had Hubbert's curve been right, 95 percent of all U.S. oil resources would be gone by 2018. Oil production in 2009 was predicted to have retreated back to the paltry rates of 1922.[31] Instead of the long bleak march downward that Hubbert foresaw, oil production in the United States has actually been on the rebound since 2008. These days, North American wells are churning out bigger year-on-year gains than any others in the world.

Hubbert also applied his method to forecast the country's natural gas production, concluding that it, too, was fated to decline in the early 1970s. On gas, he would prove even more grossly mistaken. In 2012, production in the United States climbed to an all-time high. More gas has been produced and marketed in the lower forty-eight states than Hubbert ever thought even existed in the first place.[32]

Hubbert did not stop with the United States, though; he also made bold predictions about the limits to oil production globally. According to his calculations, the world would never pump more than 33 million barrels of oil per day, or roughly one-and-a-half-times as much as was being extracted at the time. He based his claim on the assumption that the total amount of oil in the ground—the sum of everything that had been produced historically, known reserves, and projected discoveries—was 1.25 trillion barrels. Global production would hit an irrevocable peak around the time of the year 2000.

It sounds ominous, but the record ended up telling a different story. In 2012, according to the International Energy Agency, the world produced roughly 90.8 million barrels of oil a day, or 33.1 billion barrels per year, eclipsing Hubbert's figure.[33] The U.S. Geological Survey (USGS), in a 2000 study, estimates the world's total endowment at over 3 trillion barrels, almost 2.5 times what Hubbert did, though this, too, is almost certainly understated.[34] A 2008 follow-up by the USGS calculates that 565 billion barrels of conventional oil are still undiscovered. If unconventional resources, like oil shale, are also included—and Hubbert's 1956 figure purportedly included these kinds of oil in the ground as well—the figure is vastly higher. The Orinoco oil belt in Venezuela alone, laden with heavy oil, contains 513 billion barrels of oil.[35] These resources do tend to be more expensive to extract and refine than other types of oil in the ground, though. Yet improvements in technology can also lead to yesterday's "unconventional" resources becoming tomorrow's "conventional" resources. The cost of processing them tends to decline accordingly. The bottom line, contra Hubbert, remains clear: His attempt to predict the future path of worldwide oil production by arbitrarily fitting a curve to historical data has been of no real utility. Meanwhile, alarming predictions of a hard ceiling to the world's oil output have come and gone.

Hubbert's work on the exhaustion of oil supplies was seminal for a new generation of neo-Malthusian scholars in the 1990s and 2000s. Updating

Hubbert's models, these apostles of depletion argued that a downturn in world oil production was imminent, which would herald a new era of rising prices.[36] Their contention that world oil production had either reached or would reach within several years a production ceiling, which could not be exceeded due to geological limitations, was the rallying cry of Matthew Simmons and present-day "peak oil."

These arguments often get the fundamental tenets of resource economics massively wrong, though.[37] For one thing, Hubbert misinterpreted the message of the (now reversing) decline in U.S. oil supply after 1970. It was not an ominous sign of a looming shortage worldwide, only that the cost of producing oil in the continental United States was higher than importing oil from lower cost sources overseas, so oil was being drilled elsewhere. Preachers of the gospel of depletion tend to misinterpret the implication of the fact that conventional oil resources are of a fixed quantity. While that is true in a geological sense, investment expands the existing reserves and innovation in exploration and production technology tends to reduce the cost of recovering crude as time goes on. Moreover, studies that try to calculate how much oil the world can produce inevitably rely on dubious assumptions about how much oil is in the ground. But there is a tremendous amount of uncertainty about that number, which has evolved over the years.

There are other flaws. Hubbert-type depletionist arguments tend to ignore basic market mechanisms—in other words, the interaction of price with supply and demand. Rising prices discourage people from buying more oil, not to mention using it more efficiently, and simultaneously encourage people to produce more oil. Consumption does not simply grow unchecked from year to year, regardless of price. These variables are inextricably linked in real life but operate in silos in the Malthusian world. There is also a substitution effect: When the price of one good increases, people look for other, cheaper ones that can fill the same role. In the United States, for example, today's ultra-low natural gas prices are increasingly inducing power generators to use it to produce electricity, rather than coal. This process of substituting one commodity for another, driven by the search for cost savings, can help alleviate demand for a scarce good when it becomes more expensive than alternative sources.

Moreover, there can be other reasons (beyond a fixed geological ceiling) why production of a natural resource declines over time. Production of mercury, now known to be toxic, peaked worldwide some years ago—not for lack of the metal in the ground, but the market was less interested in it. In the case of oil, for example, widespread concerns in many countries around the world about the impact of oil on the environment have led public officials in countries like China and the European Union to tighten fuel economy standards for cars and trucks. Taken together, high prices, more efficient vehicles, and demographic changes have already caused oil consumption to fall in the United States over the last half–decade.

In Defense of Resource Optimism

There is a tempting intuition to the idea that the real prices of nonrenewable goods should rise, more or less, forever. It sounds right: The world's population is growing bigger by the decade, the logic goes, consuming more and more scarce resources that can never be replaced. Supply of the stuff is limited—once it is gone, it is gone. As we exhaust a given resource, it will get more and more expensive to obtain it (whether through mining, drilling, or otherwise), because the most easily obtainable supplies will be depleted. Left to exploit ever-greater quantities of ever-more-marginal deposits, prices will rise indefinitely into the future. Thus, in this line of reasoning, unless we start to consume less of a given finite substance, it will forever get more expensive.

The logic appears uncontestable at first glance. But it is wrong. The prices of raw materials have not traveled the path this story would predict in any major commodity market, once inflation is factored in, over very long stretches of time. One of the most powerfully counterintuitive and empirically conclusive findings in economic history is that the real prices of nearly all major resources have actually trended lower over very long periods of time, even if they are produced in greater volumes.[38] Although nonrenewable commodity prices can rise steeply over years or even decades when supply and demand conditions warrant, over the centuries they have tended to stagnate or decline. Betting that some commodity is "running out," which will cause its price to move ever higher in the decades ahead, has simply never been a profitable way to play the market—at least not in the last 500 years or so.[39]

This trend has been well known among economists for decades. In a landmark 1950 study, Raúl Prebisch and Hans Singer documented the tendency for the price of commodities to decline relative to manufactured goods over time, drawing on earlier work by Charles Kindleberger on this long-term deterioration in the terms of trade of raw goods.[40] Although the finding is not without its critics, the decline in real commodity prices over the very long run remains one of the most empirically grounded results that scholars have uncovered with respect to the historical behavior of primary commodities. It is also perhaps among the most surprising to nonexperts and the most misunderstood in terms of its implications.

Two useful data sources for examining historical trends in primary commodity prices are the *Economist* commodity-price index (figure 1.2) and

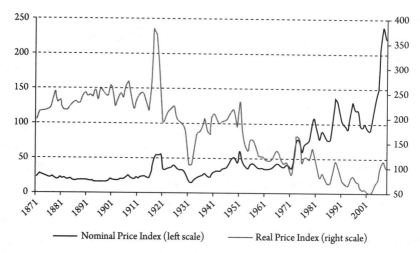

Figure 1.2. *Economist* industrial-commodity price index in real and nominal terms (1871–2010). Source: *Economist*; Shiller's 2012 historical inflation data can be accessed at http://www.econ.yale.edu/~shiller/data.htm; U.S. Bureau of Labor Statistics.

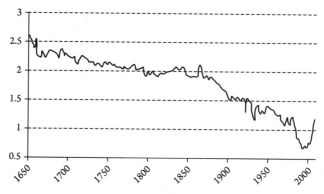

Figure 1.3. Logarithm of index of real commodity prices (1650–2005). Source: This index was constructed by Harry Bloch and David Sapsford, "Innovation, Real Primary Commodity Prices and Business Cycles," Paper presented at the 13th conference of the International Joseph A. Schumpeter Society, Aalborg, Denmark, June 21–24, 2010, see pp. 2–3. It is derived from the data compiled by David I. Harvey et al., "The Prebisch-Singer Hypothesis: Four Centuries of Evidence," *Review of Economics and Statistics* 92, no. 2 (May 2010).

another index created by four scholars, led by David Harvey of the University of Nottingham, which will be referred to as the Harvey index (figure 1.3).[41] The *Economist* industrial commodities index, first published in 1864, is widely considered to be the world's oldest public, regularly updated price index. Although the nominal index stretches back to 1845, data before 1857 is incomplete and data between 1857 and 1861 reflects January prices only. Only figures from 1862 onward, which represent averages of the underlying monthly figures, are used here. It comprises a sub-index based on industrial metals and another of non-food agricultural goods. They are deflated using U.S. consumer price index data since 1871, culled from the Case-Shiller historical home price index.[42] The industrial commodity index is a better reflection of long-term trends in nonrenewable resource prices than the all-commodity index, which also includes food prices, so this analysis focuses on the former index.[43] The commodities included in the industrials index, as well as their relative weightings, have changed over time, with the current weightings reflecting the value of world imports from 2004 to 2006.[44] The most recent rebalancing occurred in 2010. As it stands, aluminum and copper together constitute 76.8 percent of the metals portion of the index, with zinc, tin, and lead making up the remainder. Cotton and rubber each constitute roughly 30 percent of the non-food agricultural list, with timber at 14.2 percent. Wool, hides, and pressed oils filling in the rest. Crude oil, notably absent from the index, will be discussed a bit later on.

The Harvey index, created in 2010, includes the prices of twenty-five commodities stretching all the way back to 1650. To create it, data from a number of different sources of historical prices, from the U.S. Census Bureau's *Historical Statistics of the United States: Colonial Times to 1970* to Masaru Iwahashi's *A Study of the History of Price in Early Modern Japan*, had to be pooled. Prices exist for some commodities earlier than others; markets for some of them, such as crude oil, did not originate in modern form until centuries after the start of the index. Twelve of the commodity price data sets begin in the seventeenth century (beef, coal, cotton, gold, lamb, lead, rice, silver, sugar, tea, wheat, and wool), three the following century (coffee, tobacco, and pig iron), eight in the nineteenth century (aluminum, cocoa, copper, hide, nickel, crude oil, tin, and zinc), and two in the year 1900 (banana and jute). Prices are denominated in British pounds, rather than U.S. dollars, because the United States did not have its own currency until after independence. To get the most representative prices for these commodities, the average price of each one is calculated across major industrialized countries in a common currency and then added to an unweighted index. Nominal prices are deflated using an unweighted manufacturing value-added price index spliced together from various data sets covering four periods—1650 to 1784, 1784 to 1870, 1870 to 1950, and 1950 to 2004.

What does analysis of these two price series—one stretching back to 1845, the other all the way to 1650—show regarding the behavior of primary commodity prices over the very long term? Commodity prices have tended to fall over the centuries, once inflation is factored in. Raw materials prices show a secular deterioration relative to manufactured goods over very long stretches of time. Since 1871, the *Economist* industrial commodity price index has sunk to roughly half its value in real terms, experiencing average annual compound growth of −0.5 percent per year over the ensuing 140 years. Even after the boom years of the 2000s—in 2008, for instance, as commodity indexes soared, the *Economist* index never climbed more than halfway above where it stood 163 years earlier, in real terms. The message in data is the same regardless of whether they are deflated by the U.S. consumer price index or the U.S. gross domestic product (GDP) deflator, as some prefer.[45] The Harvey index shows the same general downward trajectory. Since 1650, eleven of the twenty-five individual commodities reveal a statistically significant drop in prices when adjusted for inflation (eight without allowing for structural breaks, and three more including them). Not a single one of them shows a significant upward trend in price over the last three and a half centuries. "In the very long run," conclude the creators of the index, "there is simply no statistical evidence that relative commodity prices have ever trended upward."[46]

What accounts for the downward long-run trend in real commodities prices? There are several possible explanations. First, the income elasticity of demand for most commodities is typically lower than for manufactures. As a result, increases in income tend to translate into greater demand growth for manufactures than for commodities, and hence price gains for finished goods outpace those for raw materials. Second, a long-term decline in transportation costs has had a more deflationary impact on commodities than on other goods, because these costs ordinarily comprise a higher proportion of the delivered price of commodities than of manufactured goods. Third, manufacturing has become less input-intensive over time, thanks to technological innovation, meaning relatively less demand for raw materials compared to finished goods. Fourth, productivity growth in the agricultural and mining sectors, which has been better overall than for manufactured goods, has tended to lower prices for processed goods in relative terms. For agricultural goods, productivity gains have allowed for rising global supply without the need to resort to less fertile land, while in the case of mineral goods, better ways of detecting and producing them have deepened the global reserve base and in some cases reduced production costs despite the accelerating rate at which these goods are pulled from the ground. Finally, the quality of manufactured products has improved over time. This relative shift in quality (as well as the difficulty of quantifying it, which prevents economists from isolating this variable) also may contribute to the relative price decline.[47]

The graphs in figure 1.4 of three nonrenewable resources—copper, zinc, and aluminum ore—reinforce the fact that growing production of finite

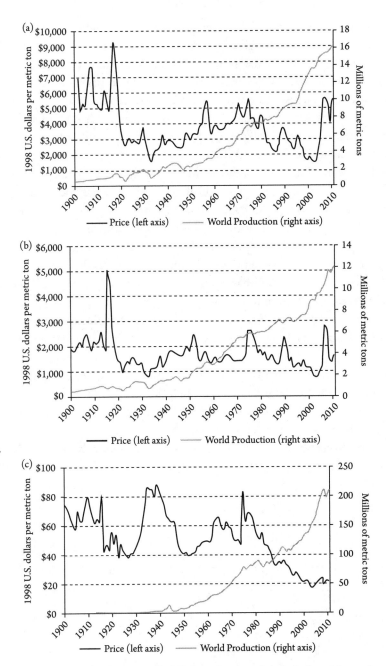

Figure 1.4. Annual real prices and global production of three industrial metals (1900–2011). Source: U.S. Geological Survey.

goods simply does not mean that their real prices are doomed to rise as they become more scarce. Although copper production has grown tenfold over the last century, inflation-adjusted prices are still lower than what they averaged between 1900 and 1920. Of course, there have been periods where real copper prices have trended upward, such as between 1930 and the mid-1970s. Occasionally prices have jumped sharply, as they did during World War I and then again during the China-centric industrialization in the mid-2000s. Still, the net effect of a hundred years' mining has been for inflation-adjusted prices to slope downward. Zinc shows a broadly similar trajectory. Production of the metal has climbed relentlessly since 1900; prices have leaped now and again, but over the long term, prices have stagnated at best or even declined. Aluminum ore, barely mined at all until the 1950s, is now extracted at a rate of more than 200 million metric tons per year. Yet prices have not followed suit. They have seen periods of massive price rises, only to come crashing down. Their unrelenting slide since 1974 has resulted in a downward net trend over the last 112 years. The bottom line is clear: Rising demand for these nonrenewable resources, reflected in greater production over time, does not necessarily mean continual real price increases, despite perennial prophecies to that effect.[48]

What about oil prices? Like these metals, crude oil is a finite resource that has been extracted in larger and larger quantities annually over the last 150 years (see figure 1.5). It took 104 years for the first 45 million barrels per day (mb/d) to be pumped, but less than half that amount of time for the rate of production to nearly double. Prices have varied enormously since the mid-nineteenth century. The real price of oil has ebbed and flowed over time

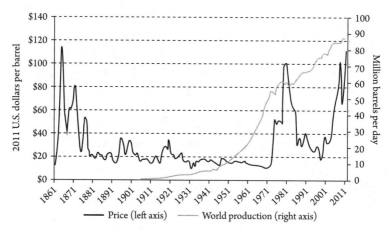

Figure 1.5. Annual real benchmark world crude oil prices and global oil production (1861–2011.) Source: *BP Statistical Review of World Energy 2012*; the Shift Project 2012.

depending on economic growth and the level of spare oil production capacity held worldwide (the buffer between how much the oil industry can pump and what it actually does). Like commodities more broadly, the real price of oil has not shown the long-term upward march that a Hotelling scarcity rent would predict. Prior to the Organization of Petroleum Exporting Countries (OPEC) flexing its muscle for the first time in the 1970s, real prices had dropped almost ceaselessly since the mid-nineteenth century. A barrel of oil was five times as expensive on average between 1861 and 1871 as it would be between 1961 and 1971.[49] Indeed, a perennial problem facing the American economy during some of these decades was that oil was too cheap—and hence a threat to the survival of U.S. oil producers, especially independent ones—rather than today's sky-high prices.

Crude oil was an exception among major commodities in that it ended the twentieth century with real prices higher than at the beginning. Why? Politically motivated constraints on oil exploration and production within OPEC countries, which reordered the structure of the global oil market, was one major reason.[50] Formed in 1960 with the primary purpose of arresting the fall in its revenue per barrel, OPEC is the most important and successful of all global commodity cartels, surviving and often thriving for several decades. Its member countries, which control the lion's share of the world's most easily exploitable oil reserves, have successfully kept their capacity far below what the market would otherwise dictate in order to keep prices artificially high. These countries' short-term means of buoying prices has been managing (somewhat unevenly) its members' production levels. Yet the arguably more effectual tool they have used to keep prices buoyant is longer term in nature: to limit foreign access to their reserves and constrain internal capacity expansions.[51] The cartel produces only 20 percent more oil today than it did in 1973, the year of the first modern oil crisis, despite oil consumption worldwide being nearly 60 percent higher.[52] Meanwhile, roaring demand growth in the developing world kept non-Middle Eastern oil companies scrambling for new sources of oil, pushing up prices and calling forth ever greater amounts of oil from the ground. Oil production is growing fastest today in places where economics dictates it should not be, like the United States, where the fields are among the most picked over in the world, largely because countries with easier-to-access resources limit their own production. Meanwhile, consumption is falling in Europe and the United States, the direct effect of a weak economy, pain at the pump, and aging populations. Yes, this is the picture of a dysfunctional market—but it is not the story of one locked in depletion-driven decline, let alone one in which inflation-adjusted prices are bound to shoot higher indefinitely. This is not the end of energy history.

New Era Economic Thinking

Economic historians have long wondered how popular beliefs and stories affect how markets behave. Robert Shiller, for example, has looked in depth at the social, cultural, and psychological forces that underpin speculative manias in the stock market and, later, the housing market. Both of those markets in recent decades experienced what many scholars of behavioral finance would agree were speculative bubbles. The dot-com craze of the 1990s saw tremendous enthusiasm for technology stocks among the American public and professional investors, who saw the popularization of the Internet as a reason for prices to zoom higher with abandon. But that boom busted when the bubble burst. Likewise, the 2000s saw a remarkable take-off in housing prices, only to collapse suddenly after the onset of the Great Recession.[53]

Looking back over the stock market bubbles that have transpired in the United States over the last century, Shiller notes that each one was accompanied by a widespread belief that fundamental changes in the economy were leading to a new era of unprecedented prosperity. The rosy developments appeared to justify the huge gains in the stock market. Each time, many experts would proclaim that prices were headed higher indefinitely into the future, perhaps even perpetually. They had trouble recognizing how closely their arguments for why the boom might last forever resembled the last time such a speculative frenzy took place, only to be proven incorrect when boom finally gave way to bust. Shiller calls this "new era economic thinking."[54]

In his conception, a "new era" is the "popular perception that the future is brighter or less uncertain than it was in the past." New era thinking has often marked "speculative market expansions" over history. Implied in the term is that such perceptions are illusory and eventually prove to have been overly optimistic. He derives the term from "the new era economy," an expression that Alan Greenspan used to describe the Internet boom of the 1990s. It was first coined in two articles in the *Boston Globe* in June 1997, along with "new era thesis," "new era theorists," and "new era school" to describe the notion that "the arrival of the Internet in the 1990s . . . was a fundamental change that would boost the productivity of the economy" as a long-term trend.[55] The term had become enough a part of the investment lexicon by August 1997 that Paul Krugman, writing in the *Harvard Business Review*, attacked the "new era theory" in criticizing equity market valuations at the time.[56] It remained in regular use until the collapse of the tech bubble in 2000, when it fell out of fashion.

New era economic thinking refers to a shift in collective opinion that the fundamentals of the economy have or are undergoing a change that will lead asset prices to continue rising steeply for the foreseeable future (and perhaps perpetually). In such a scenario, the general public "[overreacts] to the new era stories that become suddenly popular, missing the basic similarity between the latest stories and similar stories that appeared many times in the past." New era thinking "concentrates attention on the effects of events currently prominent in the news," rather than positing future alternative scenarios. New era theories emerge only after a stock market boom rather than as an entirely predictive theory of future economic behavior. That fact is important. Such theories are "after-the-fact interpretations" of market behavior. They do not originate until after the market experiencing the "new era" has already begun to show significant price advances in need of explanation. In Shiller's words, "a stock market boom is a dramatic event that calls for an equally dramatic interpretation." The advent of a new era provides such an interpretation.[57]

The market does not just react to new era theories; it plays a critical role in popularizing them. In that sense, "the stock market often creates new era theories, as reporters scramble to justify stock market price moves." A positive feedback mechanism arises between rising price levels, which spawn sweeping theories justifying the booming markets, and the theories themselves, which support further price rises and more belief in the theories. The general public is more susceptible to believing new era theories of economic growth than experts are, according to Shiller, although experts typically have a role in articulating the theory and lending credibility to it among the public. There are always high-profile individuals ready to explain financial market price movements, not to mention lower profile people interested in boosting their visibility by explaining the markets. As Shiller writes, "Whenever the market reaches a new high, public speakers, writers, and other prominent people suddenly appear, armed with explanations for the apparent optimism in the stock market. . . . Although prominent people can certainly move markets, often their wisdom merely tags along with market moves." The theories and the markets can thus feed on each other.[58]

Shiller documents how new era economic thinking has been prevalent during all four significant speculative market booms in the United States over the course of the twentieth century—1901, the 1920s, the 1950s and 1960s, and the 1990s.[59] The first of these bull runs occurred at the dawn of the century, the year 1901, which witnessed a tremendous increase in stock market prices. The new century seemed to herald a new age of high technology and futurism. Newspapers carried prophecies of what was in store: "trains will be running at 150 miles per hour . . . newspaper publishers will press the buttons and automotive machinery will do the rest . . . phonographs as salesmen will sell goods in big stores while automatic hands make change." Many of these predictions were accurate and would be fulfilled, of course. The trouble lay not in the predictions themselves, but in what their authors believed such innovations would herald for the stock market. They helped to bring about an "outburst of speculation . . . rarely paralleled in the history of speculative manias." At the time, the two most cited reasons for the market boom in 1901 were the proliferation of new technology and the growing market power of "combinations, trusts and merger"—companies that appeared to be sure-fire bets for stock pickers. The speculative fervor hit a fever pitch in 1901 and then gradually cooled until the stock market crash of 1907. The crash ended the clamor over a new era of financial returns in a major way. But the over-optimism would return less than two decades later, and with force.[60]

The bull market of the 1920s—and its crash in 1929—are the stuff of legend. The proliferation of highly public and exciting technologies, such as the automobile, rural electrification, and the expansion of radio made the era an exciting one for American consumers. It was a time of substantial real economic growth, and hence market optimism. John Moody, writing on the stock market in 1928, judged that there is "nothing now to be foreseen which can prevent the United States from enjoying an era of business prosperity which is entirely without equal in the pages of history." In 1929, Charles Amos Dice authored *New Levels in the Stock Market*, which predicted all sorts of new achievements in America—among them, a "new world of industry," a "new world of distribution," and a "new world of finance." Yale University's Irving Fisher, in August 1929, famously announced that "stock prices have reached what looks like a permanently high plateau." Dice and Fisher both had listed numerous reasons for why the seemingly new era of stock prices had arrived and would be sustained. Advances in the Federal Reserve, the rise of the holding company and growing economies of scale, improvements in investment banking, the discovery of principles of scientific management, most efficient agricultural processes—all of these developments had led to a truly new age in stock market returns. Some analysts were skeptical, but they were in the minority—at least until the crash near the end of 1929.[61]

New era economic thinking reappeared in the 1950s. World War II had fueled extraordinary growth in national production in the first half of the 1940s. When it ended, many worried that the depression of the 1930s might reappear. But a near doubling in the value of the U.S. stock market, along with solid earnings growth, between September 1953 and December 1955 put such fears to rest. The bull market fostered "the sense that investors were terribly optimistic and confident of the market was in and of itself part of the new era thinking," according to Shiller. Just as the radio had done in the 1920s, the television was transforming mass national culture, "evidence for technological progress that could not be overlooked," in his words. Inflation was mild, Baby

Boomers were increasing demand for a variety of consumer goods, and Kennedy, "perceived as showing vision and optimism," was inspiring confidence, as of 1960. "Businessmen can enjoy reasonably continuous prosperity indefinitely," proclaimed an American newspaper. And stocks appeared to be the asset class that would bring the average person that prosperity. When the Dow traded above 1,000 in 1966, it was as a national news item of the first order. But the exultant rise of the stock market would not last. No sharp correction occurred, but the period from January 1966 to January 1992 was one of low annual stock returns—just 4.1 percent.

The stock market boom of the 1990s followed closely the "new era" pattern of the previous booms. In Shiller's words, "as with all stock market booms, there were writers during the 1990s who offered new era theories to justify the market." Michael Mandel, in a 1996 article in *BusinessWeek* entitled "The Triumph of the New Economy," cited five reasons—all fundamental economic, political, and technological ones—for the boom: "increased globalization, innovation in the high-tech sector, moderating inflation, falling interest rates, and surging profits." Roger Bootle, author of the 1998 book *The Death of Inflation*, also attributed the new global "zero era" to the potent combination of privatization, capitalism, and liberal labor markets, which had permanently subdued inflationary pressures. Steven Weber's 1997 "The End of the Business Cycle" in *Foreign Affairs* foresaw a similar new era of prosperity, thanks to lower macroeconomic risks than in previous times. For Weber, the worldwide growth of the service sector vis-à-vis the manufacturing sector had led to a permanent softening of the boom-and-bust cycle of industrial output. Another common argument for permanently higher stock prices and earnings growth was that the Internet had led to and would continue to spur remarkable productivity gains. Shiller admits that "not all stories in the media in the 1990s were slanted toward new era emphases when compared with stories during earlier episodes of high pricing." The media's market optimism in the 1990s was more a "matter of background presumption" rather than the "bold assertion" it had been in 1901 or 1929. Some critics mocked what they believed were excessive valuations, but they were on the outskirts. Most people were "far more swept up in the new era thinking symbolized for them by the coming of the new millennium."

How do new era narratives come to an end? Not in one "dramatic burst," argues Shiller. The change is more gradual. Even one or two days of crashing stock markets do not cause people to up and sell once and for all. Even in the case of the 1929 crash, the market had nearly fully recovered to its previous highs by early 1930. The change in perception, rather, is "generational in character." The sharper the correction, the speedier the change in perception. When financial stocks lost more than half of their value between March and April 2000, the change in perception was rapid. If an overheated market simply plateaus, perceptions change slowly. In the case of the dot-com craze, the "fascination with the Internet stocks" went from appearing to herald a new economic age to looking like "a silly, in fact embarrassing, fad." When bearish sentiment takes hold, it can be as hard to undo as the bullish sentiment had been in the boom years just prior. In the case of the 1930s, widespread concern arose among experts that the national economic system was failing. It appeared to be entering a period, in the words of a University of Chicago professor that Shiller cites, "a stage of more or less permanent stagnation."[62]

News commentary tends to go from enthusiastic and glowing to apocalyptic. Stories of hostility among business competitors, public anger at the government and securities industry, and political strain are common. Some new economic problem can steal the market's attention. Inflation fears in the 1960s, widespread fear of overpopulation in the 1960s, and worry over the United States' loss of preeminence to Japan are examples. Convincing news stories that appear at the time the market is peaking can have a disproportionately large impact on public opinion. Such articles do not in themselves have the power to freeze the market's upward momentum, but they can begin a feedback mechanism of public opinion that spells the end of the new era narrative. Consumer confidence falls alongside the proliferation of negative stories, which then feeds back on itself, taking other economic indicators along with it. Shiller concludes that the mass switch from optimism to pessimism is, on some level, inexplicable; "market psychology somehow mysteriously changes," in his words. New era thinking ends "when the focus of the debate can no longer be so upbeat." The attention and emotional inclination of the public, once changed, is not easily, and perhaps cannot be, reversed. In Shiller's words, "there are times when an audience is highly receptive to optimistic statements and times when it is not."[63]

Epochs of New Era Economic Thinking in the Oil Market

New era economic thinking is not limited to the stock market or the housing market. This book shows that new era thinking—and all of the media craze that goes with it—has been on full display during the boom-and-bust history of the American oil market. A tightening oil market and rising prices have appeared to many experts and the broader public to herald a new era of shortage, which some predict will be permanent, or close to it. Often these experts express doubt that the rate of oil production can ever be increased, citing the ultimately limited supply of oil in the ground. Other times experts have questioned whether oil itself will soon be exhausted altogether, and with it the final collapse of modern conveniences like gasoline-fueled cars. They have fallen into the trap of extrapolating high or rising prices far out into the future, pessimistic that low prices will ever again be possible when demand appears to be relentlessly outrunning supply.

The frequency of public expressions of worry about continually rising oil prices and a looming end of oil correlate remarkably closely with oil prices. Where booming stock markets tend to generate unrealistic visions of a new era of abundance, rising oil prices do just the opposite, inspiring prophets of doom to make pessimistic statements about the future availability of natural resources—and indeed, at times, about the fate of oil-consuming nations and the industrial world more broadly. The news media carries bleak reports about impending shortages of energy and the economic threat of oil prices running unceasingly higher. These dire prognostications are not limited to uninformed people, but also take in prominent energy experts, politicians, and others. Often the oil industry speaks out against them, decrying the shortage talk as nothing more than misinformation and scaremongering. Because the public tends to be skeptical of these claims, viewing the oil industry as biased or conspiratorial, often rendering its arguments ineffective in assuaging public concerns that something may be deeply wrong.

Globally, oil production has always found a way to go higher, though the locus of production has shifted geographically over time. It has waned in some countries (like the United States) and waxed in other parts of the world. Rising oil prices have never lasted; the forces of supply and demand have always tamed them. Higher prices spur greater production by expanding the pool of oil that is economically feasible to drill. They also cause people to explore for oil in untested places, where they discover new reserves. Exploration and production technology improves, aiding the process of locating oil in the ground and lowering the cost of extracting it. Demand for oil also responds to higher prices. The more expensive oil is, the less of it people tend to consume, slowing demand growth or lowering demand altogether. They adopt other sources of energy in place of oil. Energy efficiency improves over time, allowing the same job to be done using less oil.

When prices do fall, the voices that once trumpeted the new era of higher oil prices and long-term shortage fade back into the woodwork, just as Shiller describes stock market experts doing. Just as rising oil prices tend to popularize the view that the world is facing a critical long-term oil shortage, or even running out of the stuff entirely, falling prices tend to see the shortage narrative fall out of mainstream conversation. The same newspapers that had been heralding a new era of long-term shortage only months earlier go silent on the subject as prices slip, or even crash.

This phenomenon is as old as the industry itself. A. J. Hazlett, one of the country's best known oil journalists in the early twentieth century, described it as long ago as 1918. As Hazlett wrote in the *Oil Trade Journal* in February 1918:

> At regularly recurring intervals in the quarter of a century that I have been following the ins and outs of the oil business there has always arisen the bugaboo of an approaching oil famine, with plenty of individuals ready to prove that the commercial supply of crude oil would become exhausted within a given time—usually only a few years distant.

Yet the prophecies had never been borne out:

> But always about the time it seemed as if their prophecies were about to be fulfilled, either a new pool would be discovered by some venturesome wildcatter or some one would come to the rescue with an invention that would help out wonderfully in its way. That is one reason why I cannot see any cause for alarm in the present pessimistic outpourings regarding potential petroleum possibilities.[64]

His depiction of irrational anxiety over the end of oil supplies nearly a century ago sounds and feels as though it could have been published yesterday. Peak oil may be the latest incarnation of the "new era of shortage" thinking, but swirling rumors about an imminent exhaustion of oil predate the peakists by at least a century.

Over the last hundred years, the first oil shortage narrative to gain currency in popular and political discourse in the United States lasted from roughly 1909 to 1927. The specter of a dire shortage soon to come loomed over the market, casting a shadow that one historian has since described as "singularly unreal" in retrospect.[65] Bleak assessments of domestic oil reserves from the USGS led various government leaders and geological experts to warn the public that the country's reserves were dwindling rapidly. Booming demand and fuel prices appeared to reinforce their predictions, with U.S. oil production increasing dramatically to fuel the Allied war effort in World War I and later the proliferation of automobiles on America's roads. The oil industry adamantly rejected these arguments, dismissing the fears as unwarranted, yet they persisted. Eventually, though, the tide turned. Technological advances and the discovery of vast new fields, both in the southern United States and overseas, led to a dramatic increase in crude oil production in the latter half of the 1920s, which, combined with a subsequent downward demand shock caused by the Great Depression, led to an oil price crash from their previous highs. With the market facing a crude oil glut, the oil shortage narrative quickly disappeared from the national conversation.

The next wave of new era theorizing about permanently higher oil prices took place between 1940 and 1949. After an oil glut had weighed heavily on prices for most of the 1930s, World War II proved a stunning oil demand shock. Frightening predictions of a gasoline shortage along the East Coast in 1941 never materialized, but the experience set the tone for the wartime rationing, imposed by the Petroleum Administration for War, that appeared to herald a new permanent shortage of oil in the country. With American oilfields largely fueling the Allied war effort, government and military leaders publicly fretted that domestic oil reserves would run dry in a few years' time. In most cases, the primary fear was not of a global oil supply shortage, but rather that the country's indigenous production would not be able to meet its own demand needs going forward. Still, the prospect of becoming a net importer appeared to presage a new era of higher prices and need for synthetic fuels. It was in the context of this disturbing realization—and of a dawning Cold War—that the American government began to actively assist domestic producers in acquiring acreage overseas. A brief but intense refined product shortage in 1947–48, caused by a variety of disruptions and extreme events, led to further doomsday predictions amid soaring retail fuel prices. By the close of the 1940s, however, an explosion in oil imports and production gains at home led to newfound abundance. Talk of an oil shortage fell silent when prices collapsed. The oil shortage narrative all but disappeared from American mass media for a generation.

Several forces combined in the 1970s and early 1980s to ignite intense fears about a future of permanently short supply and runaway oil prices. Soaring demand caused the spare production capacity of American oil fields to disappear by 1972. A handful of key producer countries, sensing their geostrategic importance to the West and tired of abiding by multinationals' low official prices for oil, formed the Organization of Petroleum Exporting Countries (OPEC) in 1960, eventually prompting a wave of nationalizations in the 1970s. Expert predictions of a looming "energy crisis" first appeared in 1969 and grew more common over the next few years, from the dire warnings of the U.S. State Department to the gloomy prophesies of the Club of Rome.

Amid these tensions, two oil crises during the 1970s caused government and public worry over permanent oil shortages to reach a fever pitch, the first in 1973 and later from 1979 to 1980. During this era, predictions of an imminent and irreversible shortage of oil supply were ubiquitous in government, industry, and academic circles. Surveys suggested that the American public was increasingly under the impression that the country was running out of energy. The second oil crisis of the 1970s, which stemmed from the Iranian revolution and the ensuing Iran-Iraq War, appeared to some Americans to be the fruition of the foreboding predictions made throughout the decade. Higher prices, however, tempered global demand by fostering substitution out of oil and improving conservation, while simultaneously stimulating record production outside OPEC. By the end of 1985, with its fellow OPEC member countries brazenly disregarding their agreed-to quotas, Saudi Arabia decided it had had enough of restraining its production in order to keep prices aloft. The following year, Riyadh flooded the market. With oil oversupplied, expressions of worry over whether future oil consumption needs could be met were banished for another decade and a half.

The new era narrative was reborn between 2001 and 2009. This time, the shortage viewpoint largely took the form of the so-called peak oil movement, which became an increasingly influential voice in financial markets and the energy community as oil prices rose dramatically through 2008. Oil prices climbed steadily higher over much of the 2000s, driven by Chinese and other emerging market-led demand growth, paltry production gains outside of OPEC, and a lack of new investment by OPEC in new production capacity. Commodity markets boomed. As they did, the number of books, articles, and op-eds on the topic—the world's "oil crisis," "the end of oil," "the last oil shock," or "the end of the oil age"—began to grow exponentially, with the frenzy reaching its zenith as crude oil shot up to $147 per barrel in the summer of 2008.

Discussion of the imminent, irreversible global oil shortage was not limited to the printed page but was also the subject of conferences, documentaries, and blogs. It found prominent supporters in the academy, such as Princeton geologist Kenneth Deffeyes, and on Wall Street. It found a ready audience among public officials and think tanks, with claims of an impending oil crisis issued by numerous government agencies and research groups around the world. By May 2008, oil futures for delivery as far out as 2018 were showing oil selling for more than $135 as far as the eye could see. But then the reckoning came. When crude oil prices suffered a stunning collapse in early 2009, stories of a new era of oil shortage largely fell out of fashion. Although predictions of continually rising prices did not abate entirely, the prominence and frequency they once saw was gone. Crude prices soon rebounded to triple digits, but the seeds of a moderation in oil prices, thanks to waning demand in the developed world and a North America-led supply boom, had been sown. Talk of shortage was replaced by excitement about the unfolding American oil and gas renaissance.

In *Irrational Exuberance*, Shiller mapped out the contours of new era thinking as it applies to the stock and housing markets. New era thinking in the oil market shares many of the elements that Shiller observes in equity and housing markets yet it differs in three important ways. The first of these differences is that Shiller sees new era thinking as closely associated with the formation of bubbles for assets like equities and housing. In the case of oil, though, there is no conclusive econometric evidence that irrational exuberance or excessive speculation was responsible for the general path of oil prices between 2003 and 2008, nor for the other long-term trends in oil prices over the last century.[66] The objective of this book is not to argue otherwise. Moreover, oil has only been traded in open markets like it is today for the last three decades. These markets generally function well as a means of determining fair value for a barrel of oil, and at least as well as any other market structure that has ever been devised.[67] It would be anachronistic to argue that new era thinking has led to oil market "bubbles" à la the dot-com boom prior to the 1980s, since oil-linked futures and other derivatives were not bought and sold by everyday investors on the New York Mercantile Exchange (or NYMEX) and other commodity exchanges prior to that time.

The utility of the new era concept goes beyond spotting bubbles, though. Widespread new era thinking in a given market does not necessarily mean that a speculative bubble is present. Nor does it necessarily imply strict causality between new era economic thinking and asset prices bubbles. Rather, as Shiller shows, new era stories have "tended to *accompany* the major booms in stock markets around the world" (emphasis mine). These stories are important because without reference to their details "the economic confidence of times past cannot be understood."[68] In this formulation, the existence of a widespread kind of new era economic thinking is neither a precondition nor a trigger for a speculative bubble. But it can, at the very least, help explain the social and cultural environment in which bull markets thrive.

All of this said, at certain times, popular notions of an irreversible decline in oil production have indeed appeared to some experts to play a role in leading oil prices to deviate significantly from what supply-demand fundamentals would dictate. One such anomaly arose in the spring of 2008. In mid-May, WTI crude oil for delivery the following month sold for an astronomical $128 per barrel. Normally, tight market conditions in the present correspond to lower prices for oil for delivery far out into the future (a condition known as backwardation), in part to provide incentive for oil producers and inventory holders to release it to market right away. For a stretch of time in May and June 2008, though, futures prices showed the exact opposite pattern: The further out into the future the contract, the more expensive oil was. In a May 29, 2008, report, *Oil.com*, Edward Morse of Lehman Brothers argued that growing belief in the peak oil thesis may have played a role in this anomalous behavior in the futures market:

> What has been striking about the rise in the back end of the [forward] curve over the first three weeks of the month has been—at least until last Friday—nearly a complete lack of interest by producers in selling oil forward. In the past, producer selling would have slowed or impeded the upward price drive. In its absence, retail investors, hearing a growing number of reports about "peak oil," continued to buy at higher and higher strike prices, indicating that there is no necessary physical ceiling to the market.
>
> Meanwhile, a number of banks and well-known investors publicized their views that the long run was also going to be tight for oil markets. For some of these analysts/investors, the reason for higher prices had to do with the need to trigger an adequate supply and demand response; for others, it was because they believed peak oil had arrived and, in their opinion, Saudi Arabia could no longer raise its output.[69]

It was only a matter of days after Lehman Brothers published this note that the aberration reversed itself, and the market flipped back into backwardation. If Morse's argument was correct and the peak oil narrative did contribute to irrational exuberance, it was an instance in which the narrative had left the realm of abstraction and become a potent force in the market.[70]

There is a second difference between what history shows about new era thinking in the oil market versus other markets. New era fervor in the oil market does not build until prices start to rise. But when they do rise, new era theorists often look back at past forecasts of a looming shortage on the horizon and embrace these old arguments, post-hoc. "So-and-so saw this coming," they cry, "And now the long-promised end of oil is finally upon us!" Many of the predictions of an imminent, irreversible shortage of oil that would go on to achieve notoriety years later actually preceded periods of true supply tightness by several years. Then, once supply-demand fundamentals tightened and prices rose accordingly, the predictions received more attention and a growing chorus of similar predictions and expressions of fear regarding an impending long-term, even permanent, oil shortage followed suit.

This pattern has played out several times in history. The USGS's Day Report, released in 1909, got relatively little attention when it was published. It only began to receive widespread attention in the national press, among national policymakers, and among subsequent forecasters once World War I and the car boom of the 1920s spurred a meteoric rise in the country's demand for oil and a coinciding tightening in the crude oil market. Later, Harold Ickes, Franklin D. Roosevelt's Secretary of the Interior, was pilloried in the national press in 1941 for forecasting "the beginning of a sharp and serious drop in supplies." But when oil demand boomed during World War II, his prior insistence that a severe oil shortage was on the horizon was lauded in Washington policy circles. Nearly three decades later, the Club of Rome's *Limits to Growth*, published in 1972, became a cultural touchstone thanks to the oil crises and runaway inflation of that period, which made the text appear uncannily prescient—but only for a time. Later, in the mid-1990s, a handful of analysts began to apply the statistical methods that Hubbert had used in 1956 to predict the peaking of U.S. oil production in the 1970s. Many of them determined that world oil production would enter into terminal decline between 2000 and 2010, "never to rise again." At that point, they argued, prices would begin to rise indefinitely, leading to a major crisis, unless people started consuming less oil. Their alarming conclusion appeared in major periodicals like *Nature*, *Science*, and *Scientific American*.[71] Once prices did in fact start to rise a few years later, these predictions of imminent shortage appeared to a growing chorus of analysts to be uncanny prophecy.

The third idiosyncrasy about oil-related new era economic thinking is also the most subtle. By definition, new era predictions claim that a fundamental shift has occurred in a given market, which will push prices higher "for the foreseeable future and perhaps indefinitely," as Shiller describes. For markets whose basic structure has remained constant the over last century, such as equities, it is easy to see why these sorts of claims about a fundamental, eternal turning point in the market are suspect. But the oil market is an altogether different beast. Its supply-side structure *has* undergone fundamental changes over the last century, which have led at turns to distinct, statistically discernible epochs in real prices, according to economists Eyal Dvir and Kenneth Rogoff, as figure 1.6 suggests.[72]

Two examples from history stand out. In the depths of the Great Depression, with real oil prices at all-time lows, U.S. federal and state agencies, notably the Texas Railroad Commission, began to regulate domestic production in an attempt to rein in the chronic overproduction that posed an existential threat to local economies in the oil patch. By imposing quotas on American oil producers, the U.S. government kept prices stable and sufficiently high to keep producers out of financial trouble. Thanks to this steadying hand, prices and volatility were lower on average over the next four decades than any time before or since. Four decades later, though, the world market would undergo another epochal change. In 1972, the last vestige of spare production capacity in the United States ran out, taking away Washington's ability to stabilize prices. Shortly thereafter, the newly formed OPEC, a cartel a handful of

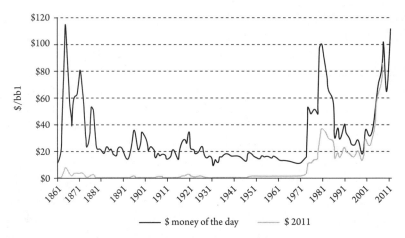

Figure 1.6. Real and nominal world crude oil prices (1861–2011). Source: *BP Statistical Review of World Energy 2012.*

countries that controlled much of the world's most cheaply exploitable sources of oil, began to use its market power to influence prices. It did so using two levers: by restricting access to most of the world's highest quality oil reserves and by rationing production within the territories of its member countries. The coalescing of OPEC as a force in the world market global supply ushered in a new epoch of oil prices emerging around 1973 that is still underway, featuring on average sharply higher and more volatile prices than in the preceding four decades.[73]

Why does this matter? It means that someone who predicted in the early 1930s that the empowerment of the Texas Railroad Commission would lead to a long-lasting period of lower average prices and volatility would have been exactly right; so too a forecaster who announced that the rise of OPEC would mean generally higher and more erratic prices than any time since the nineteenth century. In both cases, a fundamental change in the structure of the market *did* in fact lead to a new era of dramatically different price dynamics than in the decades prior. It is questionable whether long-term price formation in the U.S. equities market, as one example, has undergone a structural shift of the same magnitude over that period. Yes, stock prices have risen and fallen, sometimes dramatically or for prolonged periods of time when the right economic forces have collided. Yet no single entity in that market has enjoyed anything akin to the enduring market power of a Texas Railroad Commission or an OPEC, both capable of fundamentally altering price behavior over decades. Oil is different in that sense. Distinct eras in the world oil market, brought about by politically driven changes to crude supply patterns, have indeed transpired over the last hundred years.

The problem, though, is that experts and the broader public have often either misunderstood the implications of these structural shifts or sworn that some new-era-inducing event had occurred when none in fact had. In heat of the 1970s oil crises, for instance, the common notion among the American media was that oil prices would move irrevocably higher, driven by a new scarcity. That did not happen. Yes, world oil supply was in a new era, marked by new supply constraints—but did that mean that prices were destined to increase over the coming decades or that oil would be in permanently short supply, as many experts predicted? As it turned out, nominal prices did shoot through the roof, from $1.80 per barrel in 1970 to $36.83 in 1980, driven higher by robust global demand growth, OPEC's new proscriptions on supply, and the end of American spare capacity. But then the tables turned. Prices fell every year for the next six, collapsing to a nadir of $14.43 per barrel in 1986. And they did not stop there. Real oil prices went on to decline throughout the 1990s, from $40.83 per barrel (in 2011 dollars) in 1990 to just $17.55 in 1998.

The lesson is simple: While real oil prices have been higher on average in the forty years after 1970 than in the preceding forty years, it has been anything but the perma-bull market that new era thinkers in the crisis-ridden mid-to-late 1970s, and later in the run-up to 2008, predicted on the basis of a supposedly menacing new depletion.[74] The buy-and-hold strategy for investing in oil that neo-Malthusians would have wholeheartedly embraced in 1979—the same ones peak oil disciples would have argued for in 2008—or that Andrew Carnegie tried to execute in 1863 by digging a "lake of oil"—would have been the kind of irrational exuberance that does not end well.

When oil prices are rising, the public is often quick to chalk the trend up to depletion, or a world "running out of oil." In other words, we are bumping up against the limits of the world's resource endowment, which will push prices higher for the foreseeable future. But that explanation is facile at best and misleading at worst. The world's proven oil reserves have only increased over time. They doubled between 1980 and 2010, despite a 30 percent increase in the rate of consumption. In the United States alone, proved reserves of crude oil have risen from 2.5 billion barrels in 1899 to 30.9 billion barrels by the end of 2009, despite the vast quantities of oil that have been pumped in the meantime.[75] Global production has marched upward almost without pausing since the first oil well was drilled in the United States in 1859. Above-ground factors—like economically motivated restrictions on drilling in oil-rich countries, paltry investment in production capacity for many years (and still within OPEC), and robust consumption growth—are better able to explain today's high prices than an absolute shortage of oil in the ground. The oil market is deeply inefficient: Those countries with more than three-quarters of the world's oil reserves have only seen one-sixth the number of drilling rigs deployed.[76] There is no question that the technical cost of recovering oil—discovering, developing, and producing it—has gone up in recent years. But various ways of reckoning long-term oil prices, including comparing industry cost indices to long-dated crude futures, suggest prices are likely to stabilize at levels much lower than today's.[77] This is not the picture of a world at the end of its oil supply, let alone where prices are doomed to go up forever, despite what the irrationally anxious might believe from time to time.

"A National Crisis of the First Magnitude"
The United States Geological Survey in an Era of Booming Demand, 1909–1927

Fear of a coming oil shortage was the last thing on anyone's mind in 1901, the year that the Spindletop well in Texas—the first "gusher" in the nation's history—was drilled. Spindletop marked the start of the Texas oil boom, and with it, a new chapter in the history of oil. The epicenter of American petroleum production no longer resided in Pennsylvania, but had moved "out west," where it has remained. Rather than flowing at 50 barrels per day, as was typical of the Pennsylvanian wells familiar to American oilmen, Spindletop was capable of producing more than 75,000 barrels per day of crude oil. The newfound riches of southeastern Texas transformed the humid, rural landscape into a riotous scene of frenetic drilling and money grubbing practically overnight.

The steady growth in U.S. oil production, along with an increase in the types and consumption of refined products, defined the domestic oil scene in the closing decades of the nineteenth century and the first decade of the new one. Spindletop was just the beginning. A string of big discoveries in the Gulf Coast, Texas, and Louisiana, which yielded similarly promising new sources of supply across the southern United States over the next few years, only swamped the market farther. A succession of large finds in Oklahoma, beginning in 1901, of which the 1905 discovery of the Glenn Pool field was the most notable, would make the state the top producer until it was overtaken by Texas in 1928, a state which remains the country's most prolific. It had taken thirty-one years since the country's first oil well had been drilled in 1859 for the United States to surpass 100,000 barrels per day of crude oil production. Two decades later, by 1909, it would be just north of half a million, thanks largely to the strength of these new finds far beyond Appalachia.[1] The voluminous flow of oil from the young, booming fields of

the southwest was more than the local market could bear, and prices collapsed. But the glut would prove temporary, thanks to a rapidly growing
American population and economy that meant a swelling demand for petroleum. Between 1890 and 1910, new product categories—fuel oil, gasoline, naphtha, industrial lubricants—would soon eclipse the national appetite for kerosene, the backbone of the nineteenth-century industry.
Improved efficiency in and more sophisticated technology for producing,
transporting, and refining oil, not to mention the birth of entirely new markets, like the internal combustion engine, made these years ones of dynamic
growth for what would fast become one of the country's dominant sectors.
These massive discoveries had put the American petroleum industry on a
fundamentally altered upward trajectory.

Roosevelt Confronts the "Imminent Exhaustion" of Crude Oil

But there was another groundswell taking place, which would permanently
change the way the country related to its newfound mineral wealth: the rise
of the conservation movement as a force in national politics. "By the beginning of the twentieth century," one historian writes, "a new view of nature
appeared in America," a change that was "part of a general backlash against
the excesses and wastes of the industrial age."[2] With the closing of the American West the century prior, the sense that the country's seemingly boundless
wilderness and the riches it contained were what Americans had longed believed them to be—unchangeable, innumerable, and boundless—had faded.
The quintessentially American view of the natural world as an endless source
of plenty—vast beyond belief and bountiful beyond risk of exhaustion—was
no longer beyond doubt. The very notion of depletion—that forest, oceans,
minerals, and soil could be here one day and gone the next, and on a massive
scale—had started to enter the national lexicon. It was a concept borrowed
from nineteenth-century thinkers, like George Perkins Marsh, who argued
in 1864 in his canonical *Man and Nature* that the loss of natural resources had
caused the collapse of the classical civilizations of the Mediterranean. And
so, he warned the burgeoning American Republic, awaits a similar fate for
any modern society that wastes its natural wealth, a path he argued the nation
had already begun to travel down. With the early years of the twentieth century came growing public support for the conservationist cause, and with it,
mounting public pressure for government officials to do more to safeguard
the nation's resources. "Scientific associations and government officials
continue[d] to issue warnings," writes one historian, echoed by "a veritable
flood of magazine and newspaper articles and editorial comments pointing

out the dangers of resource depletion and the necessity of preserving our national heritage."[3]

The conservation movement reached a new prominence in Washington after the turn of the century thanks to the boundless enthusiasm of its new political champion, President Theodore Roosevelt, who made its priority a hallmark of this tenure. Roosevelt was very concerned about the "the enormous consumption . . . of resources and the threat of imminent exhaustion to some of them due to reckless and wasteful use."[4] He lamented that oil-bearing land in the eastern United States had largely fallen under the control of "large private owners" guilty of "great fraud" in their mismanagement of its abundance. For Roosevelt, the challenge of dutifully husbanding the country's natural resources was "the fundamental problem which underlies almost every other problem in our national life."[5] Having convened over a thousand national leaders at the White House in 1908, including all the state governors, Roosevelt pressed his case for the need to protect against the "imminent exhaustion" of a host of crude oil and other precious natural resources:

> I have asked you to come together now because the enormous consumption of these resources, and the threat of imminent exhaustion of some of them, due to reckless and wasteful use, calls for common effort. . . . This Nation began with the belief that its landed possessions were illimitable. . . . We began with coal fields more extensive than those of any other nation and with iron ores regarded as inexhaustible, and many experts now declare that the end of both iron and coal is in sight. . . . The time has come to inquire seriously what will happen when our forests are gone, when the coal, the oil, and the gas are exhausted.[6]

The president had made other similar statements many times in the past. "We are prone to speak of the resources of this country as inexhaustible; this is not so," he warned in his annual message to Congress in 1907.[7] In a general sense, of course, Roosevelt was right: the failure of the U.S. government at all levels up to this point had been the cause of often vast and egregious crimes against the natural landscape, the victim in some regions of excessive logging and hunting, for instance. It was a market failure he recognized and wisely advocated a role for good governance in combating. That said, the projections he loudly proclaimed about the coming end of American coal and oil were premature in 1908, to say the least, though the value of the principle he was seeking to teach—conservation of the natural environment—remains as compelling and farsighted now as it was then.

Making a convincing case for conservation to the American people, though, would inevitably require Roosevelt's administration to quantify the gravity of the problem. If the nation's natural wealth was not limitless, as the president incessantly preached and a growing contingent of the country's elites was coming to believe, then just how urgent was the need for the government to step in as a means to ameliorate any impending crisis? When it came to the country's oil reserves that task fell to Gifford Pinchot, whom Roosevelt appointed in 1905 to be the first head of the U.S. Forest Service. Now widely regarded as the father of modern forestry, Pinchot, the scion of a New England family that had made its wealth through land speculation, would become the most prominent voice for conservation, other than Roosevelt himself, during the early years of the twentieth century. It was Pinchot who convened the 1908 governors' conference at the White House that gave the president a platform for pushing his message to an audience of national officials. His utilitarian ambition—to "produce the greatest good, for the greatest number, for the longest run" through the careful husbanding of the nation's environment—was symptomatic of an era that aspired to scientific management, whether of industry or public policy, deployed by technocratic experts.[8]

Pinchot lost no time implementing the vision that he and Roosevelt shared. Out of his own pocket, he organized and hosted the Governor's Conference on Conservation in 1908, the first major national event on the subject. Its ability to draw together influential business leaders, members of Congress and Roosevelt's cabinet, Supreme Court justices, and many state governors made it a powerful platform for getting the word out to a broader public. He was soon appointed chairman of the National Conservation Commission's executive committee, assembled by the president in June 1908.[9] To impress upon his fellow leaders the seriousness of the purpose to which he was summoning them, he arranged for a reprinting of George Perkins Marsh's *Man and Nature*. Corporate control and a lack of credible data about the country's environmental resources, Pinchot instructed members of the Commission, were the enemy of sound resource stewardship. Among the small world of federal scientists, many of whom were personal friends of Pinchot, this was hardly news. They overwhelmingly shared his judgment that in governmental action alone lay the answer to the problem of resource exhaustion. But making that case nationally, and contending with the nation's powerful business interests who were unsurprisingly less amenable to Washington's growing enthusiasm for stricter regulations over land use, would be a harder sale.[10]

In the early 1900s, more was known about some aspects of the country's resource endowment than others. Forest acreage, for instance, was vastly more amenable to measurement, acre for acre, than oil or coal reserves, for example, where similar physical judgments were more guesswork than science. Wildcatters in

1909 knew little to nothing of the subsurface analytical techniques that would come later, let alone geophysical assessment, to say nothing of today's "digital oil-fields." Detailed, comprehensive well records were almost unheard of, and the lack of petroleum engineering savvy enough to interpret them meant that estimating the ultimate recoverable resource base in any given area was vulnerable to wild uncertainty. Compounding the difficulty of such estimations is that recoverable reserves, unlike farming or logging, depend on prevailing market prices. Higher prices allow oil to be brought to market that might otherwise not be at lower prices; they also encourage exploration, which results in companies booking additional reserves as they find new oil, a process that invariably becomes more accurate as upstream technology—that is, for exploration and production—improves over time. The nuanced interplay of market forces and finite natural resources, foreign to Pinchot and federal researchers of his generation unfamiliar with the economics of oil supply, would prove a stumbling block in their attempts to ascertain and forecast supply and demand conditions in the decades ahead.[11]

The Day Report

Under Pinchot's watch, the National Conservation Commission assigned the deputy to the director of the U.S. Geological Survey (USGS), David T. Day, to conduct a study of the total volume of crude oil reserves in the United States. The purpose of the report was simple: to determine how long the country's oilfields would last, assuming current trends in production and consumption hold in the future. The report was known as the Day Report.[12] The research for the Day Report was done in 1908, a time "when the agitation for the conservation of our mineral and other natural resources was at its height," one of the geologists who helped prepare it later wrote.[13] Its broad dissemination to a nontechnical audience after its publication in 1909, appearing in a Geological Survey bulletin and the *American Review of Reviews*, not to mention national newspapers, gave it a prominence that transcended Washington circles. Its message of an impending shortage of oil—scarcely comprehensible at a time when the greatest threat to the industry was too much oil, not too little—would send shockwaves across the country.

Day's conclusions were startling. There was a chance that oil production might "be increased slightly" in a few states, but other than that it was all downhill for new oil supplies in the United States. There were somewhere between 10 and 24 billion barrels left, by Day's calculation, though 15 billion seemed the "fair average estimate."[14] Approximately 2 billion barrels had already been extracted.[15] The math did not bode well for the country's future as an oil supplier: "Carrying out the increasing rate of production," and given declining output from fields already in production, "the industry would be brought to an abrupt end by exhaustion, except in California, by about 1920," Day wrote in the *American Review of Reviews*. His forecast as published by the USGS the same year was a bit more optimistic—domestic oil reserves would be exhausted by "about 1935."[16] Day acknowledged "the possibility and even the probability of the discovery of other oil fields as explorations are extended," but the figures in the report "must be considered as representing our actual knowledge of the oil resources of the United States."[17] Yes, there was some uncertainty around the exact timing. Yet Day's number-crunching had "yielded fairly accurate information concerning the known oil supplies of the country."[18] The bottom line, from his perspective, was clear:

> The estimates of different authorities will vary between wide limits, but they will all agree that the known fields are being exhausted at a rate so rapid as to mean cessation of the industry within a few decades unless the expected new fields are found, and this reliance upon unknown sources of supply after a few decades seems to be the characteristic attitude, as if these new fields of great size were a foregone conclusion.[19]

Optimism that more oil would be discovered would only be naïve. Thus, "with the certainty of exhaustion of the present fields by the present generation, it is not a matter of vital argument whether such exhaustion comes in ten years or forty."[20] In other words, the end of American oil was nigh—it was only a matter of exactly how soon it would come.[21]

Based on his bleak forecast, Day urged Americans to take action. "In the face of an approaching scarcity," he wrote, "the use of petroleum should be limited to the purposes for which it is essential and for which no other material can be substituted." Lighting houses without access to gas or electricity counted, as did the "still more essential use" of oil for the purposes of industrial lubrication. Cutting down on waste was essential. "Waste," as Day meant it, constituted any use where another energy source, like coal, could do just as well. Oil had no place as a fuel for locomotives or power generation, as was currently taking place. But the greatest waste of all was in exporting crude oil and refined products to other countries. To remedy these abuses, he advocated that the federal government should immediately cease to sell any public lands "where petroleum is probable." Furthermore, the Geological Survey should conduct a careful investigation into the exact geography of oil-bearing land so that any remaining acres yet unknown will not be put on the market. In other words, Uncle Sam, not wildcatters in Texas, was best equipped to make decisions about where and when oil resources should be tapped. These measures "should be initiated at once," Day's report concluded, "to be of benefit when the present petroleum supply becomes inadequate."[22]

Day's conclusions were carried in the Report of the National Conservation Commission, published in February 1909. In the preface to the report, President Roosevelt gave a hearty endorsement to its findings with regard to the startling shortage of oil. He called the report "one of the most fundamentally important documents ever laid before the American people," which "deserves, and should have, the widest possible distribution among the people." Although he conceded that its conclusions were only "an approximation to the actual facts" given the "knowledge and time available," he urged U.S. officials not to "defer action until complete accuracy in the estimates can be reached, because before that time many of our resources will be practically gone. It is not necessary that this inventory should be exact in every minute detail." He reminded Americans that "our mineral resources are limited in quantity and cannot be increased or reproduced." The end of the oil age in the United States was not far distant, barring unforeseen improvements in technology or finding ways to curtail demand. "With the rapidly increasing rate of consumption" of the nation's "mineral resources," such as oil, "the supply will be exhausted while yet the nation is in its infancy unless better methods are devised or substitutes are found."[23]

In reality, though, the Day Report was an imperfect study on a number of counts. Two historians, Diana Davids Olien and Roger M. Olien, pull no punches in enumerating the methodological weaknesses of the report. The study simplistically assumed a single national average porosity of oil plays, average yield per cubic foot of pay, average thickness of pay for all fields, and average rate of recovery for all fields. Such assumptions "in operational terms amount to nonsense" and are further undermined by the poor quality geological data on which the report was based, due partly to how quickly the research was thrown together. Furthermore, the Day Report did not evidence any awareness of the effect of oil prices on recoverable reserve estimates or future production. It assumes one fixed rate of future extraction regardless of the market price of the good and fails to acknowledge potential future fluctuations in the country's consumption of oil. On top of all these shortcomings, Day's study did not view any future additions to petroleum reserves as likely enough to be factored into the supply equation. It does not take into account the potential for finding additional oilfields, improving recovery rates at existing oilfields or adding to them. Yet its conclusion was unmistakable: American oil would be completely exhausted in perhaps as few as eleven years.[24]

The cataclysmic conclusions about the future of the American oil industry contained in Day's report would receive wide airing in the two decades following its release. Its bleak portrait, apparently amply supported by the best geologic sleuthing in the country, became the bedrock on which follow-on studies with similar objectives would build. Olien and Olien describe its central place in the national dialogue on energy supply:

> Few conservationist perspectives on petroleum have ever received as much attention and repetition as Day's. For the better part of the next two decades, the USGS repeated his alarm in the petroleum section of its annual survey of mineral resources; Day wrote this section of the report until 1915 and was followed by John D. Northrop and David White. His successors kept up the same campaign with the same ammunition.[25]

Part of that publicity can be attributed to the dogged efforts of George Otis Smith, whose lengthy tenure as the director of the USGS stretched to 1930.

Smith followed up Day's 1909 report with an article that May in the *Annals of the American Academy of Political and Social Science* touting the accuracy of his agency's findings, lest anyone second-guess them:

> An inventory of the mineral resources of the nation recently made public by the National Conservation Commission owes its chief value to the fact that the data used had been in preparation for more than a score of years. . . . This bureau [the USGS] in its explorations and investigations, had accumulated quantitative data that became readily available at the time of popular awakening to the needs of national conservation. . . . Herein lies the practical value of much of purely geologic study. Thus, an important part of the Survey's work through these years has been to keep the country informed as to the occurrence of economic minerals.[26]

Science had come to the aid of the policymaking process, helping to put the heated debate over what would become of the country's oil wealth on technically sound footing. So Day's study was not the only such warning to the American people. Rather, its importance lay in the fact that it had the explicit backing of the federal government, which lent its findings an air of authority, while its apparently rigorous methodology suggested scientific credibility to its gloomy prognosis. Smith reminded readers that hard work may yet prevent U.S. energy supplies from driving into the ditch, though it was by no means a sure thing:

> In the view of the rapidly increasing demand upon these supplies it also becomes imperative that strenuous effort should be made to

discover new sources of mineral fuels and ores. Thus alone can optimism be justified. Only as geologic explorations and mining operations uncover new deposits and block out known reserves can the United States with the continuance of its industrial development avoid facing the sure exhaustion of the supply of certain important minerals within this century.[27]

The multiple re-publications of the Day Report, including in the popular *American Review of Reviews*, followed by forecasts like Smith's that drew from the same well, advanced the perception among the broader public that oil was on a near-term road to ruin in the years leading up to the World War I.

Rumors of an impending oil shortage buzzed in newspapers in the years following Day's report, as they had in those leading up to it. An aside in a December 1912 newspaper article noted that the supply fears apparently in the air that year "recalls that of five years ago when word was passed about the land that the wells were running out." Spurred by these stories, "the price went up. When it reached the top notch oil men went into Texas, Kansas, Louisiana, and California, sunk wells, struck plenty of oil, sold it at big profit and the price dropped back."[28] In July 1910, a reporter for the *Hartford Courant* interviewing J. J. C. Clarke, the "publicity man" for Standard Oil, noted "the daily papers" carried "persistent reports" that oil "is in danger of becoming exhausted." Clarke's answer was hardly reassuring: "Although I cannot give you the exact figures at this instant, I am inclined to believe that the supply will hardly be equal to the enormous demand. It certainly will not exceed it to any great extent."[29] Such flat talk was uncharacteristic of the oil industry in this era of booming production, though it was commonplace among other energy experts. The geologists of the USGS were not the only men of learning prophesying doom as the decade wore on, as a 1913 article in the *Christian Science Monitor* makes clear:

> Professor V. Lewes delivered a lecture recently at the Royal Society of Arts on "Liquid Fuel," in which he dealt with the problem arising from the increased consumption of oil. The demand for oil exceeded the rate of production, and a shortage seemed inevitable. It was doubtful if in 50 years' time oil would be obtainable in sufficient quantities, and at a price that would enable it to be used commercially. There were grumblings with regard to rings and combines, but the simple explanation was that demand was getting ahead of supply.[30]

Shortage was in the air, apparently the stuff of rumors and speculation in the years surrounding the release of the Day report in 1909.

"There Is Nothing to Fear"

But Professor Lewes's remarks also reveal a more subtle public perception: that the oil companies, led by Rockefeller's Standard Oil, were using foul play to push up the price of petroleum for their own gain. The dark hand of "rings and combines" was a public enemy more than familiar to Americans in this era of Roosevelt trust-busting. "No general question of governmental policy," wrote framers of the soon-to-be Interstate Commerce Act of 1887, "occupies at this time so prominent a place in the thoughts of the people as that of controlling the steady growth and extending influence of corporate power and of regulating its relation to the public"—and Rockefeller's Standard Oil was public enemy number one.[31] The company was the undisputed goliath among its peers in the first decade of the twentieth century, controlling as much as 91 percent of oil production and 85 percent of final sales in 1904.[32] Even after 1911, when the U.S. Supreme Court upheld a court order demanding the Standard Oil group be carved into pieces, the oil industry was hardly an object of goodwill in the public's eyes. When markets tightened, the leading vertically integrated American oil companies were never above suspicion of being responsible. The mistrust that characterized public opinion of Rockefeller and his Standard Oil Company has largely survived up to the present day, though the objects of suspicion—the major oil companies and the structure of the oil market—have certainly evolved.

Standard and its heir companies did enjoy a sway over U.S. oil prices in the years during this era, including after the court-imposed dissolution of Rockefeller's behemoth. A glimpse at how prices were set during this era illustrates just how differently it operated from the modern market, where oil-backed securities are traded openly on commodity exchanges and physical oil changes hands in relatively deep spot markets. Beginning with the creation of the Titusville Oil Exchange in 1871, crude oil was traded through oil certificates on various exchanges throughout the country's producing areas, which were limited to a wide swath of land from Indiana to New York. Trading on the exchange was "wildly speculative" through the 1880s, but as the amount of oil traded on the exchanges gradually declined, Standard decided to flex its muscle when it came to how much its pipelines and refineries would pay producers. In 1895, Standard Oil, through its buying arm, the Joseph Seep Agency, announced that it would name its own price for oil—the "posted price" of petroleum. In effect, Standard would now have a say over the price of the lion's share of the crude oil produced in Pennsylvania and Ohio, which it would set "as high as the market of the world will justify." It defended the arrangement to be only natural, as it had "before us daily the best information obtainable from all the world's markets," which it drew from to make "the best possible consensus of prices."[33]

The role of Standard Oil in setting prices for both crude and refined products from the 1880s to 1900 was virtually unchallenged. Even after the dissolution of the Standard Trust in 1911, the former Standard companies continued to play a dominant role in setting downstream prices for the entire industry. "Prices announced by the Standard companies are generally accepted as the market price," said a U.S. Federal Trade Commission Report in 1922.[34] Citing Standard's posted price for crude as the contract price in their own transactions was common practice among American oil companies. Particularly when it came to adjusting prices in response to changes in fundamental supply-demand forces, the former Standard companies held sway.[35] Rather than encroach on each other's geographical spheres of influence at the retail level, these companies allowed their peer companies to act as uncontested price leaders within their own respective regional domains.[36]

The oil industry was by no means unified in how it responded to the rampant shortage predictions of the mid-1910s, but typically gave them short shrift. More or less complete dismissals of such forecasts sometimes found their way into the *Oil and Gas Journal*, a leading trade publication, during these years, which provides a glimpse into what those in the industry were saying to one another. "The fear that there may not be a sufficient supply of oil for transportation and industrial purposes is groundless," the *Journal* wrote. "Even if there is a manifest falling off in the eastern fields of the United States"—which looked increasingly likely—"one has only to remember the vast fields of Mexico, with their wonderful supply of petroleum ... to see that there is nothing to fear." The "search for new oil fields," which was "going on all over the world," would no doubt strike vast new reserves. Any claim to the contrary was one of the "many objectionable matters" that could be "charged up to the conservationist movement."[37] What any of them said publicly stemmed as much from their individual business strategies, which varied widely, and from their efforts to read the political tea leaves. Only the largest of the oil companies, like the former Standard companies, appear to have tailored their public statements in light of what rewards a law changed or a statute altered might bring them.[38]

Oil executives tended to dismiss top-down analyses of the country's reserves that were at odds with their own experience and intuition. A February 1916 article in the *Oil and Gas Journal* lamented the numerous "prophets who say that in 20 years known oil fields are to be quite generally drained and an oil famine of the genuine sort will prevail." "Practical oil men," who "expect the oil supply to last much longer than a score of years," could not be fooled by forecasts that amounted to no more than "fanciful guess work."

"To set a time when oil shall cease to be a marketable commodity," the *Journal* wrote, "was foolhardy." "Nobody knows how much oil is yet to be found in the earth. That there shall be an exhaustion of the supply some day is a reasonable and logical conclusion. . . . But it will require long periods of time to remove from the earth its vast stores of mineral wealth" and likely come only after oil reaches prices "never expected in bygone times."[39]

Oil in a Time of War

The advent of World War I in 1914 brought a new dimension and heightened urgency to discussion in Washington about whether American oil supplies could be counted on in the future. Petroleum was no longer just another commercial item, of interest merely to those seeking the new-found thrill of a "joy ride" or household illumination. It was now a strategic commodity. At the urging of Winston Churchill, then First Lord of the Admiralty, Britain had suited its navy to gasoline instead of coal in 1913, increasing oil's centrality to the war effort.[40] The United States quickly followed suit. Military leaders knew that this war would be won on the back of mechanized warfare. That meant that the products wrought from petroleum—naphtha, gasoline, and diesel, not to mention industrial lubricants—were crucial for the very survival of the patrimony.

Concern among senior U.S. officials over how the military would fuel itself in the years ahead was enough to prompt William Jennings Bryan, the Secretary of State, to plead with President Woodrow Wilson in 1914 to send the troops into Veracruz, Mexico, to restore order in its oilfields. The imminent decline in domestic oil fields, he reasoned, meant that occupying part of Mexico was crucial for more than just safeguarding the assets of U.S. companies in that part of the world. Rather, the imminent end of American oilfields made those in Mexico "the inevitable source from which, in the near future, the supply of oil for the United States Navy will largely be drawn."[41] Getting them working again, and firmly under the control of U.S. oil companies, was a national security issue, not merely an economic one. Wilson authorized the occupation, but soon withdrew U.S. forces after encountering unexpectedly fierce opposition.

Not everyone agreed with this about-face from Mexico. Mark Requa, a distinguished commercial geologist who acted as an advisor to the Department of the Interior, saw the United States taking control of Mexico's oil as a vital bulwark against a shortage-driven apocalypse: "It is doubtful sources of supply now unknown can be developed in the United States to compensate over any long period of time for the decline of known fields," Requa wrote in a February 1916 report to the U.S. Senate. "It is highly improbable, in fact, that a quantity equal to the present estimated reserves"—no more than 24.5 billion, he argued, citing the Day Report—"can ever be developed in territory yet to be discovered." (Proved reserves in the United States amount to 30.4 billion barrels today, despite the fact that more than 200 billion barrels of oil have been pumped since 1859).[42] In foreign oil alone lay energy salvation:

> We must either plan for the future or we must pass into a condition of commercial vassalage, in time of peace relying on some foreign country for the petroleum wherewith to lubricate the highways of commerce, in time of war at the mercy of the enemy who may either control the source of supply or the means of transportation; in either event . . . our country will resound to the martial tread of a triumphant foe.

Requa's fear was not so much of a total absence of oil the world over, but rather of a United States unable to supply its own needs. "When it is too late," he warned, "we will awake to the fact that the oil resources of the world are in foreign hands, and that, so far as its lubricants are concerned, the United States has become the vassal of some foreign power." For Requa, having "no assured source of [new] domestic supply in sight" inevitably meant that "the United States is confronted with a national crisis of the first magnitude." Insofar as the United States was headed to becoming a net oil importer, Requa's prediction was right on the money. For him, having to import oil to quench its demand would put the country in an unacceptably vulnerable position relative to other, potentially antagonistic, nations. It was a view that would win many backers, particularly among the country's military elite, during the World Wars and the early days of the Cold War—and one that still resonates today.[43]

It is not surprising that Washington was struggling to get its head around the implications of the astounding growth in domestic oil consumption during World War I. Although the world's leading producer of crude oil during World War I, the U.S. oil industry strained to supply the Allies' needs. By 1914, the country produced more than 266 million barrels, amounting to 65 percent of the world total; by 1917, that figure would increase to 335 million barrels, or 67 percent.[44] At one point, the United States was providing a full 80 percent of Britain's wartime oil imports.[45] As the war stretched on, European leaders grew increasingly panicked over a scarcity of fuel. Gasoline is "as vital as blood in the coming battles," wrote a fretting Clemenceau to Woodrow Wilson. "A failure in the supply of gasoline would cause the immediate paralysis of our armies" and could very well lead to "a peace unfavorable to the Allies."[46] Despite the full awareness of the need for more oil supplies in Washington and London, though, ever-escalating demands for petroleum products continued to push production capacity to its limit. The age of the automobile had arrived, forever transforming the use to which petroleum would be put in the United States. Given the soaring demand for oil in conjunction with the war effort, though, this massive new draw on the nation's oil production could scarcely have come at a worse time. All in all, worldwide consumption of petroleum would increase by 50 percent between 1914 and 1918.[47]

A rarity just a few years earlier, more than a million cars had been licensed in the United States by 1914, with annual production running at over half a million. That figure rose to 1.7 million cars just five years later.[48] The sales figures for Ford's legendary Model T, which launched a popular revolution in American travel by combining an attractive value for the dollar with a versatile, reliable machine, speaks to the extent of the car boom. Just over 10,600 were produced in 1909. By 1916 that number had exploded to more than half a million. More than 10 million would be sold by 1924.[49] This proliferation of cars placed tremendous new demand on oil production—no longer for kerosene, as had been the case in the nineteenth century, but for motor gasoline. In 1899, virtually all of the 156 thousand barrels per day of crude oil produced domestically were converted into industrial products, mostly as cleaning solvents and heating. But by 1914, motor gasoline made up 40 percent of retail petroleum products sold in the United States. Gasoline, about 25 percent of the value of total U.S. refinery runs in 1909, had shot to 85 percent by 1919. The demands of the war, coupled with the proliferation of automobiles on America's roads, were the primary demand-side factors that pushed the price of gasoline from 15 cents in 1914 to a high of 24 cents in 1918.[50]

A Dangerous Situation

Meanwhile, reports of an imminent peaking of American oil production—and the end of global oil supplies—kept coming. Ralph Arnold, a USGS colleague of David Day, who had authored the famous 1909 report on dwindling oil reserves, had more bad news for the country in 1916. In the Annual Report of the Smithsonian Institution (though first published in an academic journal, *Economic Geology*, in 1915), Arnold reported that the country had pumped about 3.3 billion barrels since 1857—more than 50 percent of its total remaining resource of 5.8 billion barrels. Two hundred and sixty-five million barrels had been produced in 1914 alone. Only 5.8 billion barrels were left. State by state, he estimated how much oil remained in the ground and how much had already been exhausted.

Texas: 33 percent of its total oil reserves already drained; California, 24 percent; Oklahoma, 67 percent; and so on. At the current rate of production, all of the oil reserves in the United States would only last another twenty-two years. Having already peaked, the country's oil production was doomed to "gradually decrease from year to year," dwindling to a trickle over a period of from fifty to seventy-five years.[51] By his math, there would not a drop left by 1990 at the latest.

Some commentators noticed the irony between how much crude oil was being produced and the omnipresent fears that there soon would not be enough. One attributed such attitudes to rising prices, which appeared intractable. "Despite the fact that never before in the history of the United States has so much oil been produced as today," wrote a reporter for the *Los Angeles Times*, "during the last decade the increasing price of oil . . . has created what many gas men regard as a 'dangerous' situation."[52]

The death of domestic oil reserves was not a matter of if or even of when, but rather of short-term, calculable inevitability, according to a 1916 Senate inquiry. By its reckoning, American oilfields still contained some 7 billion barrels. But "most of the American oil-fields had already reached and passed their prime and were on the down-grade." Given past trends in consumption growth, the country's oil reserves would likely last no more than twenty-five years, "according to the most optimistic calculation." So the dreaded date of the death of American oil would be 1941, in other words, if not before.[53] According to A. C. McLaughlin, an "oil expert" who testified at the hearing, the shortage of oil looming for the United States might become nothing short of "embarrassing," unless production "was greatly increased," a scenario of which he was not particularly hopeful.[54]

Two scientists sponsored by the Smithsonian Institute, Chester Gilbert and Joseph E. Pogue, were no more optimistic, fearing oil prices would rise indefinitely as the fields ran dry. Although they admitted that coming up with an exact number of how much oil remained underground was not easy, they did not suffer from a surfeit of humility about their powers of estimation, either. "While unmined petroleum, like other mineral resources not exposed to sight, cannot be inventoried with a nicety of exactness," they wrote, "the proven and prospective oil fields of the United States are, nevertheless, so broadly known that the petroleum reserves may be estimated within a very reasonable margin of error." They warned that consumption was increasing so rapidly that a "continuation of the present tendency would exhaust the petroleum remaining in an alarmingly short period of time." Yes, they admitted, more oil perhaps may be found, and yet "the most generous allowance of margin to cover possible underestimates of future discoveries does not materially change the nature of the issue." The bottom line was indisputable: "A big fraction of the domestic petroleum is gone," and "there is no hope that new fields, unaccounted in our inventory, may be discovered of sufficient magnitude to modify seriously the estimates given. [The war] has merely brought into the immediate present an issue underway and scheduled to arrive in a few years." The federal government should do all it can, they argued, to curtail wasteful production, encourage development of oil shale resources in the Rocky Mountains (the energy source to which the U.S. military would soon have to turn for more domestic oil), and promote the use of ethanol instead of gasoline.

All of this was bad news for oil prices, they claimed. Now that the peak of U.S. production was in sight, prices were bound to rise as the cost of producing oil rose, posing a serious threat to the economy. "Just so soon as the aggregate output is compounded of senile and youthful fields, with the latter no longer in the ascendency," they argued, "the resource as a whole will pass inevitably into a period of slowing and more costly production, even though the resource is yet but half exhausted." This inevitable decline in production rates, marked by higher and higher costs associated with extracting oil, meant a new "period of economic stress" due to this oil-driven "concatenation of circumstances," which would come to pass far before the actual "physical exhaustion" of the country's reserves.[55]

There was no doubt that the final years of the war had placed tremendous strain on the nation's available capacity for pumping oil. The stress on the market became increasingly visible. In 1917, imports from Mexico had to fill the orders for fuel oil and gasoline that domestic producers could not. Discoveries of new fields lagged, notwithstanding a record number of wells being drilled in the late 1910s, driven by a doubling in crude oil prices between 1914 and 1918. Even the weather was uncooperative. An unusually cold winter in 1917 and 1918 raised coal demand for heating homes. Oil, a substitute product, saw increased demand as coal prices increased in the face of shortage. Total domestic oil consumption swelled by roughly a quarter between 1916 and 1918. Consumption had overtaken production plus net imports by late 1917, which required drawdown in commercial crude stocks, a major cause of concern among brass in Washington in charge of regulating the fuel market to ensure enough of the stuff for the troops. Oil prices were jumping almost across the board. An index of wholesale prices for all crude and product types rose by nearly 100 percent from 1915 and 1918.[56]

Rumors began to surface in early 1918 that Washington might soon impose gasoline rations, in a desperate attempt to help ameliorate the shortage. Few major news outlets saw them as credible. "Nothing is easier to get into circulation than a scare," wrote a *Los Angeles Times* dispatch in January. "Considerable concern has been aroused among automobile dealers over the rumor that the government would be compelled to establish restrictions on gasoline consumption. . . . Automobile dealers already harassed by rumors assumed looks of gloom." In fact, "there is ample supply of gasoline. Inquiry at authoritative sources proves this and automobile owners throughout the country can rest assured of this fact." An executive of the Packard Motor Car Company had little patience for those who read into the stock draw a dubious future for U.S. oil supplies in the years ahead. "Not the slightest basis in fact exists for the rumor that a shortage of gasoline menaces the motor industry or the American way of life."[57] A similar appeal for calm appeared in the *Christian Science Monitor* later that year. "The recent announcement that there was a scarcity, with worse to come, was not accompanied by any explanation of the reason for this situation, and there are oil experts who simply think that there is no basis for such an assumption. One of these experts said that this is the time of the year when gasoline stocks always run low. . . . The statement, therefore, that there is a shortage in the supply of gasoline probably means that the stocks are running low, as they always do this time of year."[58]

Not every shortage prediction during the war was about long-term supplies or prices. Much of the worry that filled the airwaves and newsprint in the closing years of World War I was about short-term shortages, driven by temporary outages, infrastructure bottlenecks, and insufficient capacity to pump or refine more oil. Anxious public statements by the Federal Fuel Administration often stoked these types of concerns, though more so for coal than oil. Founded in August 1917 by executive order, the newly formed wartime agency was tasked with regulating all aspects of the nation's fuel markets, including coal, natural gas, and oil-based fuels, until it was disbanded shortly after the war ended. One of the ways it tried to tamp down civilian consumption of resources with vital military applications, like coal and oil, was by loudly reminding Americans that these goods were precious, finite, and troublingly understocked. They also sent posters to every mechanic and fuel station in the country, instructing American drivers about five ways they could prevent waste. Although the harsh winter of 1917–18 had passed, and with it the extra demand for oil driven by inadequate coal inventories, some of the Fuel Administration's leaders still doubted that the worst of the coal shortages had passed. In an advertisement urging Americans to conserve coal, P. B. Noyes, the director the bureau of conservation for the Federal Fuel Administration, viewed the country's worsening scarcity of the commodity a "catastrophe" and "silent tragedy," as it risked undermining the production of steel needed for the war effort in Europe.[59]

Coal was not the only energy source where production was bumping up against capacity. Although many the year prior had doubted he would take such dreaded measures, President Wilson, observing the strain in the oil market as the war dragged on, imposing "Gasoline-less Sundays" beginning in August 1918. Although it lasted only six Sundays, the mandated rationing drove home the fact that oil demand was exceeding ready supply. The shortage situation was even worse in Britain, and reports of its still more stringent rationing policy filled news reports in the United States.[60] When it came to the future of American oil supplies, "Everything conspired against optimism," in the words of one historian.[61] At least one trade journal for farmers worried aloud whether there would be enough gasoline to run farm equipment, a relatively new problem in an era where mechanized agriculture was fast becoming the norm. "The country [is] facing a crisis such as it has never faced before," wrote one Midwestern farmer's digest, which it predicted might lead to widespread famine.[62] Some brass from the Fuel Administration warned against such worries. W. Champlin Robinson, in a speech in Washington, reassured drivers that "supplies of gasoline are ample to take care of our war and normal requirements if we will practice sane conservation," according to a write-up in *Motor Age*, a weekly magazine.[63]

Not everyone was convinced. Some saw in these temporary shortcomings in domestic oil output a country—and a world, no less—where production would very soon enter terminal decline due to depletion, perhaps not more than a few more years hence. Edward Acheson, a one-time employee of Thomas Edison and a renowned chemist, was one of them. Writing in *Forum*, a well-regarded magazine of the day often featuring some of the country's most acclaimed thinkers, Acheson had a stern warning for his countrymen. "A rather gloomy writing is appearing on the walls of the petroleum industry." Oil, the stuff of civilization, may soon run out altogether. He put his alarming message in stark terms:

> The importance of oil to the civilized world can best be grasped when we consider that the clothing we wear, all the articles on our person, the houses we live in and practically every article in our houses would not exist as we now have them were it not for lubrication. Picture to yourself our great cities, New York, Chicago, Philadelphia, Boston, and others. . . . Think for a moment what would occur to these great cities should the world run short of lubricating oil.

Yet running short of lubricating oil, he argued, was exactly what was about to occur—permanently. Acheson said he was optimistic enough to believe that "undoubtedly many other deposits of petroleum will be found in the world," though none blessed with as much as the United States. Notwithstanding whatever other oil may be found in the future, consumption was simply overrunning geology's ability to keep pumping oil at 1919 levels. "Our best advised authorities agree that so few as sixteen years may see the practical exhaustion of the

United States' petroleum, and, further, they do not look forward to any foreign country surpassing in productivity that of the United States." And then the amusing line: "And still I am not a pessimist; far be it from me," but "I admit that if we judge solely by appearances the prospects do not look very bright and rosy for our children."[64]

Looking at the astounding increase in oil prices over the last half-decade, the average American would have had little reason to doubt that oil supplies were, in fact, running short. Intuition would have told them that oil, a finite resource, was bound to get more and more expensive over time. Short-term trends in the market certainly would have seemed to lend credence to that perception. By the close of the 1910s, refiners were scrambling to get their hands on enough crude to feed the rising demand for valuable products like gasoline. Oklahoma refiners, who just a few years earlier were feasting on the flow from the giant Cushing field, could not get their hands on enough crude oil to fully utilize capacity. Some obtained only enough to run at 50 percent of capacity. Their bidding war for raw crude shot the regional price per barrel up from $1.20 in 1916 to $3.36 in 1920.[65] In Pennsylvania, the price of crude rose from $1.75 in 1914 to $4 per barrel by autumn of 1918, and in Illinois from $1.12 to $2.42 per barrel over the same period.[66] This aggravated situation in the market likely did little to appease any sense among experts and everyday Americans that the oil supply situation might indeed be entering a new phase of permanent peril. Prophecies of a lasting shortage hung in the air, and the situation on the ground appeared to bear them out.

"The Effects of Fifty Years of Negligence … Are Now Becoming Visible"

The same passion that had gripped Requa in 1916—a vision of the United States using its rapidly expanding military and diplomatic muscle to help U.S. oil companies secure oil supplies abroad—did not abate with the armistice that ended World War I in November 1918. American oil company executives began lobbying increasingly hard for acquiring foreign oil concessions, presumably with the help of Washington, as a way of meeting burgeoning domestic demand for petroleum products. They appeared to fear that if new sources of supply were not brought on stream immediately to bring down prices, oil could lose its competitive position vis-à-vis coal and demand for gasoline and lubricating oil could be destroyed. Today, the thought of American oil companies operating abroad is second nature. But it was not always that way. The United States was by far the dominant oil producer in the 1910s, pumping as much as 70 percent of the world's oil that decade. It was the country where everyone else wanted to come and drill. Washington chafed at the thought of having to go abroad for oil, hat in hand, as Britain was obliged to.

Yet, due to the exponential growth of worldwide consumption, going abroad appeared the best way to keep reserve growth coming. In the words of one historian:

> It came as a bitter realization for American oil companies to find that by 1919 the British companies, which were currently turning out less than five percent of the world's production, had somehow acquired more than half the world's estimated future reserves. With the alarm over the depletion of American's domestic supply, the long-range future looked bleak to oil men, who saw that they might have to forego profits because of their failure to acquire abundant foreign reserves.[67]

Sir E. Mackay Edgar, a British oil executive, gloated in *Sperling's Magazine* in a widely circulated 1919 article that the Americans had foolishly run their oil reserves all but dry, giving London a new competitive advantage in the race for industrial supremacy. "America has recklessly and in sixty years run through a legacy, that, properly conserved, should have lasted her for at least a century and a half. But the effects of fifty years of negligence and inefficiency are now becoming visible." For Britain, Edgar argued, this was wonderful news. "America is running through her stores of domestic oil and is obliged to look abroad for futures reserves," which "are very largely in British hands or controlled by British capital. Before very long, America will have to come to us for the petroleum she needs." Therein lay Britain's opportunity to squeeze its cross-Atlantic neighbor as it struggled in vain to acquire foreign oilfields already under British control. "We hold in our hands, then, the secure control of the future of the world's oil supply."[68]

The dreaded specter of a foreign oil monopoly, brought on by the imminent depletion of U.S. reserves, was enough to spur Washington to action. One of the Wilson administration's ideas for offsetting Britain's feared dominance of maritime trade was the creation of a federal merchant marine fleet. At the president's request, the U.S. Shipping Board, which had overseen merchant shipping during wartime, would continue with the construction of the vast fleet it had begun during World War I, with the aim of solidifying the country's place as a leading world commercial power. Edward Hurley, the director of the Shipping Board, worried that the country—and indeed, the entire world—may not contain enough oil to power this massive new fleet. Hurley's engineers had calculated the fuel requirements of the fleet, once completed, based on deadweight tonnage. Although they admitted the data informing their calculations were "not very ample," his team calculated that fueling a global merchant fleet could soon require as much as 80 percent of U.S. oil production, and half of the entire world's production.[69] In a world where oil production had already peaked, the rules of the game by which great powers sought to outdo each other in war and in commerce had changed dramatically. For Hurley, it was unclear if the world would ever be capable of producing enough oil to power its needs in the years ahead. But one thing was certain: the United States needed to do all it could to get its hands on whatever foreign oil reserves it could.

Mark Requa, who had advocated such an effort since the occupation of Veracruz in 1914, had been urging the industry to acquire overseas acreage to prevent the United States from being an also-ran in the new global hunt. Appointed as the director of the Oil Division of the wartime Fuel Administration in January 1918, Requa and his strategy gained momentum. It was he who had warned Congress three years earlier that the country had "officially been put on notice" over a national crisis stemming from a shortage of oil reserves. The solution, as Requa saw it, was for the federal government to support American oil producers in their effort to obtain overseas oil concessions. In a letter to the head of the U.S. Fuels Administration, H. A. Garfield, Requa and two colleagues, USGS Director George Otis Smith and Bureau of Mines Director Van H. Manning, urged that "American interests be encouraged by sympathetic Governmental cooperation in acquiring additional sources of foreign supply and by protection of properties already acquired." To defend its growing merchant fleet and expand American claims to foreign oil abroad, the cadre of technocrats called for "a world-wide exploration, development and producing company financed with American capital, guided by American engineering, and supervised in its international relations by the United States Government."[70] The mainstream American oil industry agreed with Requa and his government colleagues that additional supplies did need to be brought to market if the industry was to avoid destroying demand for its products by letting prices remain at unheard-of levels. Where there was disagreement was over the percentage of future domestic demand that domestic supply was capable of sustaining at affordable prices. American crude oil production had not been able to meet domestic demand since before World War I, with imports making up the relatively small shortfall.

The former Standard companies knew that gaining control over as much good oil acreage overseas as they could was good business regardless of how long they believed domestic fields would last, especially since cheap foreign oil sent to the United States could in theory undercut their market share. Even though one oil executive acknowledged that "the bulk of oil men do not give any of these estimates [of dwindling reserves] serious consideration," they were more than willing to leverage predictions of shortage if such ideas would impel the federal government to support their overseas endeavors. A. C. Bedford, an executive for Jersey Standard, made his case for the American

government to intervene on behalf of companies like Jersey by noting that the
USGS had, after all, estimated that over 40 percent of the nation's oil was already used up. For companies like Jersey Standard, government backing for
foreign exploration could only be a good thing. Even British oil executives, in
their rush to acquire foreign petroleum, did not really believe the United States
was running out of oil. "I don't expect you or any other oil man in American
really believes that your supplies are going to be exhausted in the next 20 or 30
years," wrote the director of Britain's *Petroleum Executive* to an American
colleague.[71]

Even after observing the skyrocketing oil demand of the late 1910s, American oil executives simply did not see a peak of U.S. production anytime soon.
Industry leaders rarely addressed the issue directly, but when they did, they
typically offered a flat dismissal of the notion that oil reserves would soon be
exhausted, either globally or domestically, as careful research by Diana
Davids Olien and Roger M. Olien of the University of Texas shows.[72] "There
is plenty of petroleum and always will be," Harry Sinclair said in a speech to
the American Petroleum Institute's annual meeting in 1921. "Exhaustion of
the world's supply is a bugaboo. In my opinion, it has no place in practical
discussion." Thomas O'Donnell, the first president of the American Petroleum Institute and a pioneer of the California oil industry, expressed frustration that "the public has frequently been alarmed by statements of well-mean-
ing and learned scientists, predicting an early exhaustion of our petroleum
resources." In truth, he argued, domestic production would continue to provide at least some portion of the country's needs "long after the time limit set
for exhaustion by some of our experts." H. G. James, the president of the Mis-
souri Jobbers Association, did not mince words in telling the *Oil and Gas Journal* in April 1920 what he thought of those who argued that world oil production had already peaked:

> I am wholly out of sympathy with those croakers who are constantly
> keeping the public mind inflamed with dismal predictions of declin-
> ing production and nearby exhaustion of the supply of petroleum. The
> surprising thing is that some oil men engage in the same sort of bunk
> or are persuaded to approve what is being said by others. . . . There
> never was a time when so many people who do not know anything
> about oil were giving expert testimony thereon.

But arguably the most penetrating assessment came from A. J. Hazlett, one of
the country's best known oil journalists in the early twentieth century. For
Hazlett, such shortage fears were cyclical. The hysteria currently in vogue was
just the latest swing of the pendulum:

At regularly recurring intervals in the quarter of a century that I have been following the ins and outs of the oil business there has always arisen the bugaboo of an approaching oil famine, with plenty of individuals ready to prove that the commercial supply of crude oil would become exhausted within a given time—usually only a few years distant.

Yet the prophecies had never been borne out, Hazlett explained. "Always about the time it seemed as if their prophecies were about to be fulfilled, either a new pool would be discovered by some venturesome wildcatter or someone would come to the rescue with an invention that would help out wonderfully in its way. That is one reason why I cannot see any cause for alarm in the present pessimistic outpourings regarding potential petroleum possibilities."[73] Such a statement from one of the oil industry's leading lights underscores just how common these fears were in this era. They also illustrate just how skeptical those actually in the industry, as opposed to researchers and analysts within the federal government, were about the accuracy of these predictions.

As more and more Americans took to the roads in cars in the 1920s, the price of retail gasoline began to assume an importance to households across the country that earlier might have thought little about the market. Car ownership was multiplying at a scale hard to conceive of. A total of 3.4 million automobiles registered in the United States in 1916 had swelled to 23.1 million automobiles by 1920. Not only were there more cars on the road, they were being driven ever-longer distances. The average car traveled 4,500 miles in 1919 but nearly double that by 1929. The entire orientation of the oil industry shifted as gasoline took center stage as the end product of crude oil people wanted most. Drive-in gasoline stations sprung up overnight, from 12,000 in 1921 to 143,000 in 1929. More than four times as much gasoline was consumed during the final year of the 1920s than had been used a decade earlier.[74] Crude was as expensive in 1920 as it had been at any time in the prior two decades. In 2012 dollars, the price of a barrel of oil spiked from $14 in 1915 to nearly $35 in 1920. It would collapse the following year before resuming its path upward, this time stopping at around $20 per barrel in 1927, just before a massive new discovery in Oklahoma cause it to fall sharply again. Regardless of these fluctuations, the average American was feeling the cost of gasoline in the pocketbook as he never had before, as more and more of it went to filling up the tank. The future of crude oil supply, which in the era of Roosevelt had been more of an academic, or perhaps moral, concern, became an increasingly home-economic one.

Oil "Must Pass Its Peak at an Early Date—Probably within Five Years"

With oil prices as high in 1920 in real terms as they had been in the two decades prior, public worry over oil depletion was in the air. "There has been true anxiety in the public mind over the fuel question," went a report in the *Wall Street Journal* in October 1920, "in some cases bordering on despair. There have been legitimate grounds for such apprehensions, and doubtless it was greatly accentuated by those who find their harvest in the extremity of others."[75] As usual, these stories were buttressed by in-depth articles in leading journals and magazines about whether the world's oil reserves would last long enough into the future to be a viable energy source for the rising generation. And as usual, such reports drew on projections from studies and figures put forward by the federal government. Albert G. Robinson, a distinguished former reporter for the *New York Evening Post*, laid out the case for the imminent death of the oil age in a piece for *Outlook*:

> It is not impossible, and perhaps not improbable, that children now living will in their old age tell their grandchildren of . . . an evil-smelling liquid called gasoline used for the propulsion of automobiles. As conditions of our daily life, we are prone to regard coal and gasoline very much as we regard air and water, as things that, because they now are, therefore must always have been and must always be. As a result of scientific and official surveys, we are told that our coal supply will last for many generations. The estimations of duration vary, but none threatens any early exhaustion. But the oil situation is quite different. Within the last few weeks the Director of the Bureau of Mines predicted that "in less than twenty years the supply still underground will be exhausted."[76]

In other words, the irrevocable verdict of geology found that oil was doomed to be the fuel of the past, not of the future—and by 1940, no less.

David White, the chief geologist of the USGS, had told Americans in no uncertain terms in May 1920 that the days of American oil production were numbered. The country's oil production would hit its peak in 1927 and decline thereafter, he and his team concluded. Given White's status in the U.S. scientific community, especially in Washington, this was no small proclamation. Peak oil had arrived in the United States. It was "very doubtful" the country would ever be able to pump more than an "improbable maximum" of 1.2 million barrels of oil a day (it had reached 1.0 million the year prior):

> The production of natural petroleum in the United States must pass its peak at an early date—probably within five years and possibly within three years—though the long sagging production curve may be carried out beyond the century. . . . Very probably within seven years, the production of this country [will] pass its climax.

According to White, the numbers were all straightforward, and there was relatively little room to disagree with this official view of the USGS. He dismissed out of hand the notion that any changes in the price of oil or improvements in technology for finding and producing it would alter the picture much. Experts at the Survey had "conservatively estimated" that 6.7 billion barrels of oil remained in the ground, of which 43 percent had already been produced. He conceded that this calculation was "of necessity highly speculative," noting that "for many areas of the country the information is very fragmentary and inadequate." Nevertheless, he and his fellow scientists had relied on "voluminous" data to justify more than a little certainty about and been "sufficiently proven" to justify more than a little certainty about their conclusions. "It is highly improbable that the error" in their estimate of total U.S. reserves "is more than 50 percent. An error of 75 percent seems so improbable as not to justify serious consideration at present." What oil lay beneath the rest of the world was far harder to say, wrote White. Underscoring his point was a world map of producing and potential oil reserves, published by the Survey in 1919, notable for the large question marks placed over "Persia and Mesopotamia"— what are now the Arabian Peninsula, Iraq, and Iran.[77]

Not everyone was convinced that a crude shortage was imminent, however. Skeptics of the shortage were just as fervent in their public statements as those who sometimes sought to allay such fears. J. L. Jenkins, automobile editor of the *Chicago Daily Tribune*, derided "Uncle Sam's highly disturbing gasoline famine" as "purely mental." The "gasoline panic" was nothing more than a "specter which has swept the country, which caused no little amount of financial loss and anxiety in automotive circles generally. It has been purely psychological." A rising tide of supplies meant the "world can now emerge safely from the cloud of gloom perpetrated by the premature prophets who for months have been drawing word pictures of a gasless universe."[78] Some voices in the industry tended to look at such scares as a cyclical phenomenon. Howard Pew, the son of an oilman who would grow to become president of Sun Oil (now Sunoco), reflected in 1925: "My father was one of the pioneers in the oil industry. Periodically ever since I was a small boy, there has been an agitation predicting an oil shortage, and always in the succeeding years the production has been greater than ever before."[79] A fellow oil executive, Walter Teagle of

Standard Oil of New Jersey, regretted that "pessimism over crude supplies had become a chronic malady in the oil business."[80]

"Embarrassing Surpluses"

A few years into the 1920s unprecedented volumes of oil had begun to rain down onto the market. Old debates over shortage continued as the onrush of crude silently quickened, though almost unnoticed at first to those outside the industry. Crude production doubled to 2 million barrels a day between 1919 and 1923. It would near 3 million by 1929. Industry holdings of crude oil exploded from 136 million barrels in 1920 to 400 million barrels four short years later—a massive amount that has never been seen again in the eight decades since.[81] As one historian of the oil market summed up the situation, "Throughout the 1920s, but particularly after 1924, as public officials, industry executives, and the man-on-the-street continued to speculate over whether or not the United States would soon run out of crude oil supplies, there was so much crude oil floating around in the States that such speculation seemed premature to say the least."[82] If press reports were any indication, the shortage specter still hung in the air, despite this growing glut of oil.

"The United States is face to face with a near shortage in petroleum supplies so serious in character," blared a *Los Angeles Times* report in September 1923, "that it threatens the very economic fabric of the nation. . . . Unless immediate steps are taken to remedy this situation, the industrial, economic, and political existence of the United States soon will face one of the most serious crises in American history."[83] The following year, President Coolidge, trying to finally stem the tide of overproduction, created the Federal Oil Conservation Board, entrusting it with "safeguard[ing] the national security through the conservation of our oil."[84] So much oil was coming out of the ground that oil companies were having a hard time containing it, stopping losses from evaporation, and preventing environmental run-off. Conservation again was the watchword for Washington—but the central challenge was dealing with too much, not too little, oil flowing out of the ground.

A rapid succession of new technologies and discoveries was rapidly adding to the pace at which crude oil was being pulled out of the ground and refined products churning out of refineries. Forward strides in oil geophysics and other technological advances transformed the way oil was located, extracted, and purified. No longer would oil hunters need to rely on instinct alone to find oil, thanks to a host of techniques and tools developed for that purpose. The birth and popularization of Anticline Theory—the realization that the specific gravities of oil, natural gas, and water can lead to them becoming trapped in what look more or less like underground hills, detectable from the surface—made it vastly easier for oil hunters by the mid-1910s to map oil below ground. Beginning in the early 1920s, stratigraphy, or subsurface analysis, which scrutinized the results from exploratory wells to get a more precise understanding of where oil was located, also spread throughout the industry. Seismic prospecting, first used by the oil industry in Eastern Europe, arrived in U.S. oilfields in 1923. It proved wildly adept, its early trophies including the Greater Seminole field in Oklahoma, which peaked at 527,000 barrels a day in 1927.

There were other advances as well. Aerial surveillance, which national militaries had put to use in tracking army movements during World War I, was another means of better identifying promising geological formations, although from above ground. The magnetometer, first utilized during this era in oil prospecting, helped locate oil by measuring changes in the vertical components of the earth's magnetic field.[85] Drilling also improved markedly during the 1920s. In 1918, the deepest wells had a range of 6,000 feet. By 1930, that figure had reached 10,000 feet.[86] The biggest breakthrough downstream was the growing use of thermal cracking, a refining process first developed in 1913. Cracking enabled refiners to extract higher volumes of lighter, more valuable products, like gasoline, from the same amount of crude oil. Thermal cracking allowed refiners greater freedom to produce a variety of final products in greater quantities.[87] Technology on the demand side also made forward strides. Automotive fuel efficiency increased, in keeping with consumers' incessant demand for lighter cars that would enable them to receive the maximum gasoline mileage possible.

More oil than ever before was coming onto the global market, even as global demand was about to spiral downward. Oil discoveries flourished in the second half of the 1920s. The discovery of the giant Yates field in Texas soon established the Permian basin in the southwestern United States as one of the world's leading oil-producing regions. Oklahoma boomed. The output of the Greater Seminole field was extraordinary, growing from 192,500 barrels per day in January 1927 to over 490,000 barrels per day by July.[88] The discovery of the East Texas oil field in 1930, the largest American oilfield ever found outside Alaska, led to production at the previously unimaginable rate of 900,000 barrels per day—when the total U.S. output was 2.4 million barrels per day. Domestic production had done what many thought impossible, doubling every eight years between 1912 and 1928, when it hit 2.5 million barrels per day on average.

Foreign oil production also made new highs. Soviet oil production, which had nearly stopped as a result of the country's political turmoil, came back in the 1920s. Production as low as 75,000 barrels per day in 1918 hit 275,000 barrels per day by 1929. Yet Mexico would surpass Russia in 1921, producing more than 100,000 barrels per day. Production in Venezuela soared from just 19,000 barrels per day in 1919 to more than 500,000 barrels per day ten years later. The South American nation would become the world's largest exporter of crude by 1939.[89]

When the financial crisis of 1929 and the early 1930s broke out, it was simply too much for the oil market to bear. Demand for petroleum crumpled around the world. Along with the price of oil, the force behind the imminent shortage scare collapsed to a large extent under the weight of the new discoveries and drastically diminished demand. Prices were a shadow of what they had been a decade earlier. A barrel of oil in the U.S. Midcontinent had cost $3.50 in 1920. By time of the crash in 1929, the price was only a fraction of that amount, at $1.45. Once the calamity of the Depression for consumption had a chance to kick in, and as U.S. oilfields continued to produce with hitherto unknown potency, oil prices were cut in half again, now at $0.69 per barrel.[90] The oil world of World War I had been turned upside down. A 1926 article published in the *Los Angeles Times*, which poked fun at the "Jeremiahs who periodically predict this fatal exhaustion" of oil, had become prophetic in the space of a few years. "Perhaps we should not take these alarmist announcements of oil exhaustion too seriously," it opined. "Perhaps, in spite of [the] predictions, we shall still be building new reservoirs to take care of the embarrassing surpluses, while someone else is assuring us that 'exhaustion' is at hand."[91]

By the time prices crashed, the oil shortage narrative of the 1910s and 1920s, which had gone by a few names—the "oil exhaustion" and the "gasoline famine," among others—largely fell out of mainstream conversation, at least as far as press coverage indicates. The bleak assessments of domestic oil reserves from the USGS, undertaken in the context of the nascent American environmental movement, had led various government leaders and geological experts to warn the public that the country's reserves were dwindling rapidly. Booming demand and fuel prices had appeared to reinforce their predictions, with U.S. oil production increasing dramatically to fuel the Allied war effort in World War I, and later the proliferation of automobiles on America's roads.

Technological advances and the discovery of vast new fields, both in the southern United States and overseas, led to a dramatic increase in crude oil production in the latter half of the 1920s, which, combined with a subsequent downward demand shock caused by the Great Depression, led to an oil price crash from their previous highs. With the market facing a crude oil glut, the oil shortage narrative quickly disappeared from the national conversation. Washington would have a new problem when it came to oil—namely, how to keep prices high enough to save the domestic industry from the overwhelming glut it had trapped itself in during the early 1930s.

"A New Era of Scarcity and Higher Prices"
Wartime Demand and the End of American Self-Reliance in Oil, 1940–1949

As the 1940s dawned, neither the oil industry nor Washington had any idea how radically the dominant theme in the oil markets the decade prior—overabundance—would contrast with the preoccupation with shortage in the decade to come. Too much oil, rather than too little, had haunted the American oil patch after the demand collapse of the Great Depression. Estimates of domestic reserves, which exploded in the late 1920s and 1930s, had made the madness of the previous years' oil panics look increasingly wrongheaded. Finding new oil across the United States turned out not to be a problem. Between 1935 and 1939, new discoveries exceeded current production at an average rate of 1.3 billion barrels per year. During the first half of the 1940s, the pace of that buildup would fall to just 0.4 billion barrels annually.[1] The magnitude of discoveries in East Texas alone, which had been enough to flood the market in 1929, had forced a newly contrite industry to acquiesce to a system of production quotas overseen by the Texas Railroad Commission and other federal and state regulators. Such public policies and industry regulations were designed to prevent the kind of price collapse that had ensued during the lean years of the 1930s. They did the trick. The price of a barrel of oil in the Oklahoma-Kansas region, which had fallen to almost one-third of its value between 1926 and 1932, regained much of its strength by 1934. Crude prices along the Gulf Coast had also been pummeled—$1.45 per barrel in 1929 to just $0.80 by 1931. But they too bounced back, up to $1.29 per barrel by 1937.[2]

"The Situation Is Perilous"

By 1940, however, Washington was starting to come to grips with the threat of a new challenge facing the world's oil supplies: the threat of another global war.

World War I had etched in the minds of military leaders around the globe the vital importance of petroleum products like diesel to any modern war effort. Oil was no longer just another commodity, or for that matter, just another energy source. It was a strategic commodity. Access to oil was a matter of national survival. Ready supplies at the assembly line and near the locus of battle were an issue of national security.

Yet, notwithstanding the potential of a war-driven demand for oil to overwhelm the capacity of what existing fields could produce, U.S. government officials were generally supremely confident that the country's newfound oil abundance could produce enough to fuel any war effort. As Robert E. Wilson, an oil expert at the Office of Production Management, told Congress in April 1941: "The production branch of the petroleum industry is today able to meet any increase in demand which may conceivably be placed upon it." He acknowledged the "obviously large demands of mechanized war and the difficulties this country experienced in meeting the great increase in demand during the last war." But the country's proven resources were at an all-time high, he argued. Domestic production alone could meet consumption demands without resorting to imposing rationing measures of any kind.[3] Increasing production was just a matter of "opening the valves," in the words of one expert. After all, as U.S. oil experts proudly noted, the country had pumped 1.4 times more oil in 1941 than it had in the first forty-two years of the industry's history.[4] The cushion between current rates of production and existing production capacity kept confidence high among American officers about the country's oil situation going into an uncertain future.

Ordinary Americans first felt the full effects of World War II on the energy markets in the summer of 1941, even before the country had entered the conflict. In early 1941, the British tanker fleet responsible for transporting petroleum products from the East Coast of the United States across the Atlantic had suffered heavy losses inflicted by German U-boats. President Franklin D. Roosevelt, responding to London's cries for assistance, loaned 50 American tankers to the British in an effort to reinforce their fleet. He would loan fifty more over the next six months.[5] The tankers were assigned away from shuttling refined products from the Gulf Coast to the Eastern seaboard, a region that consumed a full 40 percent of domestic petroleum production. Fewer tankers servicing this vital route from the shipping hubs of the Gulf to the urban demand centers in the Northeast meant a bottleneck waiting to happen.

By the spring of 1941, Harold Ickes, the Secretary of the Interior who would serve as Roosevelt's point man on oil issues throughout the war, began to worry that a supply crunch was looming. Ickes was not a popular figure. Ridiculed by one senator as a "common scold puffed up by high office," Ickes had gained a reputation as a sort of conservationist prophet of doom.[6] A long-time member of Roosevelt's New Deal staff, he had frequently warned Americans of the bleak future that lay ahead unless they reformed their gluttonous ways and began to conserve oil and other exhaustible resources. In the words of one historian, Ickes was "representative of a new alarmism," a man who had made his name, as well as many enemies, by raising his voice about the ecological unsustainability of the American lifestyle.[7] He was capable of going to extremes in his cause, once threatening to arrest any driver who wasted oil by making "jack-rabbit starts."[8] Ickes was so unpopular among Republicans that the party's nominee for president in 1944, Thomas Dewey campaigned on the promise of firing him if he were voted into office.

Ickes, who chaired the Office of Petroleum Coordinator for National Defense (or OPC, which would later become part of the Petroleum Administration for War), was not optimistic about the short-staffed U.S. maritime fleet being able to supply the urban Northeast. Yet he and his team also knew that putting in place a substitute means of transporting the oil would be impossible anytime soon. Expanding overland transportation by pipelines, rail, or truck would take years to carry out, and expanding the American tanker fleet to make up for the vessels on loan to the British would take years as well. Since the supply problem appeared intractable, the OPC considered how it might mitigate the other side of the oil market equation: demand. But this would not be easy, since Ickes lacked the authority to put in place a formal rationing program. The administration opted to partner with the oil industry to urge the American public to consume less voluntarily—never an easy sell. In the autumn of 1941, a barrage of paper and radio ads in the Northeast urged people to lessen their use of all oil-based products for the good of Europe's war. Oil companies even handed out stickers for drivers to put on their car windshields, "I'm Using One-Third Less Gas." It was a rare instance in which an industry implored consumers to buy less of its own products. And Americans would have none of it. Their lack of enthusiasm stemmed in part because they "could not understand why the statements of abundance a few months before should suddenly be reversed to a claim of shortage."[9]

For the six months prior, Ickes had warned newspapers that the White House would likely need to institute rations on electricity and oil to prevent an oil shortage from occurring before the end of the year.[10] When his summertime call to the public to voluntarily conserve gasoline went unheeded, he pushed for stronger measures. By August 1941, the OPC called for deliveries of gasoline to gas stations and other outlets to be gradually trimmed down. Deliveries at every level of distribution were decreased by 10 percent effective immediately, then by an additional 5 percent relative to pre-war levels in September. The agency had no formal power to implement consumer rationing, but it did have substantial sway over the industry, which it leveraged by issuing formal wartime "recommendations" to oil companies.[11]

These new rations were puzzling to an American public that had grown used to the idea that the country had plenty of oil, and the unpleasant memory of so-called gasoline-less Sundays still lingered. Local officials, particularly on the East Coast, did what they could to assuage their constituents' fears. This oil crisis would be temporary, U.S. officials reassured an anxious nation that depended on cheap, plentiful oil to heat their homes and drive their cars. "It is not my intention in any way to give any statement which will in the slightest degree alarm anyone," one local Connecticut mayor told his constituents.[12] But alarm was exactly the effect of Washington's new rationing regime. That was no surprise, really, given the language with which Ickes and his federal oil administrators often described the "perilous" situation in the oil market:

> The Eastern oil shortage has finally reached the acute stage. . . . We must conserve gasoline stocks now to avoid a drastic shortage this winter when our tankers must be used to haul fuel oils unless our people are to be left to freeze and our defense industry shut down for lack of power. This is the beginning of the sharp and serious drop in supplies that we forecast. The situation is perilous.[13]

For most Americans, the prospect of too little gasoline, which meant they had to drive less, was annoying, but the threat of too little heating oil—the stuff many of them depended on to heat their homes in the winter—was more alarming. The starkness with which the OPC described the looming crisis did not exactly instill confidence that the situation was under control.

Some officials were merciless in their criticism of how Ickes and the OPC were handling the purported East Coast gasoline shortage. Congressional leaders from the Northeast called for an investigation in response to their constituents' anger over the austerity measures in the fall of 1941. Roosevelt's opponents were adamant that no actual shortage existed, going so far as to accuse his administration of bad faith in fanning the flames of a shortage scare that summer. They argued that Washington's top energy bureaucrats were responsible for the inaccurate shortage warnings and needless rationing measures, which were part of a broader attempt to scare the American people into the war (exactly how these shortages would have helped do that is unclear). As Representative Hamilton Fish of New York put it, "I have reached the definite conclusion that the shortage of oil in the east is phony and is being used as part of a program to create further war hysteria."[14] Senator Francis Maloney implied that the Roosevelt Administration may be spreading the scare "for psychological reasons," in other words, to prepare the American public for an unpopular entry into the war in Europe.[15]

Other officials, including congressional investigators, did not see bad faith, but rather an overly ambitious federal government that had gotten in over its head in trying to dictate domestic oil consumption. Senator David I. Walsh of Massachusetts called the shortage predictions "a glaring example of bureaucratic blundering," implying incompetence, but not deceit. One journalist chided Ickes for "demonstrating how not to deal with the increasing shortages of materials that are certain to make themselves felt in the future," succeeding only in "frighten[ing] the wits out of the public."[16] Congressional investigators, after hearing all the evidence, were likewise unsympathetic to how the Roosevelt Administration had handled the whole affair: "We are of the opinion that this unnecessary alarm was caused by over-enthusiasm on the part of those charged with the direction of the petroleum situation. There is no gasoline shortage in the East."[17]

Ickes and his colleagues did not go down without a fight, however. They accused the congressmen who contradicted their shortage reports as being "near-sighted prophets" who were putting America in harm's way. Deputy Commissioner Ralph Davies of the OPC went so far as to assert that "those who attempt to create confusion and dissension regarding the gasoline situation"—by denying that it never existed—were guilty of "sinister and planned sabotage."[18] But the investigation's harsh verdict, that the oil shortage was nothing more than a mirage, had done its damage. "To many, Ickes was just shouting wolf again," in the words of one historian.[19] It severely undercut the case that Ickes and Davies were trying to put forward, of a country that needed to conserve its oil during the war or else face potentially dire consequences. It was a fatal blow to public belief in the reality of the shortage, and consequently, public support for the rationing.

In reality, the loan of the U.S. tankers to Britain had caused a bottleneck in the U.S. oil supply infrastructure at an inopportune moment, when civilian demand for oil as well as military consumption was climbing. British oil inventories were rapidly declining, so there was little room for Washington to ask for the tankers to be sent back to U.S. waters. It was only thanks to a series of aggressive measures by the country's wartime energy ministers that adequate transportation between the Gulf and New England was accomplished. Tanker shipments to the Caribbean were put on hold indefinitely. The westward flow of oil products from the Eastern states to the Gulf of Mexico was curtailed. Most importantly of all, railway tank cars, which had been taken out of service for shipping oil in favor of more cost-effective pipelines and tankers, were pressed into service at an astounding scale. By October, oil transported by rail to the East Coast climbed to forty-five times what it had been a year earlier.[20]

But the details of the situation were mostly lost on the popular press. A host of conspiracy theories were aired across the country about who was behind the apparently misleading shortage rumors. Speculation abounded about the cause of scare. Some reports claimed that the true cause of the impending shortage was that United States was running out of crude oil once and for all.[21] Other groups, such as the New York Oil Heating Association, saw in the forecasted shortage a conspiracy by the oil industry to "build up a panic among the fuel oil distributors and the public" in order to push up prices and gain exemption from antitrust laws for the duration of the war in Europe.[22] Oil heater vendors had good reason to be upset at talk of an oil shortage, since the prospect of oil being more expensive or unavailable would likely lead to a loss in market share to coal-based heaters. James A. Moffett, chairman of the California-Texas Oil Company, held a special press conference to present his case that the government's claims of shortage were nothing less than a dishonest and thinly veiled attempt to take over the oil industry entirely, once and for all. Ickes was constructing a "smokescreen," as he called it, behind which he was "hiding efforts to disrupt the industry," thereby tarnishing its reputation in an attempt to win public support for his long-sought goal of nationalization.[23] Skeptics rallied behind a claim that the railroad industry had 20,000 idle tank cars ready to solve the transportation bottleneck, but that the government refused their offer to put them in service. Like many doves in Congress, they believed the government was stirring up the shortage fear as a means of conditioning the public for an unpopular entrance into the war.[24]

By the latter half of 1941, the kinks in the Allies wartime oil supply chain began to get worked out. Things did not turn out as bad as some in Washington had feared, though a crisis had only narrowly been averted. Britain returned twenty-five of the seventy-five tankers it had borrowed throughout the summer to their American operators. Rail shipments of oil from across the country to the East Coast increased from a mere 5,000 barrels per day in June to 141,300 per day by the fall.[25] The construction and implementation of additional oil pipelines from the western and southern United States also helped to compensate for the drop in maritime oil delivery. The demand side of the oil equation also eased, as a result of the civilian rationing program in place during October and November, as well as the restrictions on gasoline deliveries to retail hubs. Once Ickes was satisfied that the dreaded shortage would not occur, and after a tremendous public outcry, the War Production Board's delivery rationing order was revoked on November 30.

"A Have-Not Nation in Oil"

The United States had pledged to serve as the "arsenal of democracy," as Roosevelt put it, for the Allies in World War II. And yet the sheer scale of the undertaking caught the nation off guard, requiring enormous investment in new infrastructure and record gains in industrial output and efficiency. The industrial capacity of the nation was severely tested. Of all the challenges that American businesses faced during the war, the oil industry's assignment to provide for the Allies' astronomical energy demands was among the greatest. All in all, the United States would supply 6 billion of the 7 billion barrels of oil consumed by the Allies between 1941 and 1945.[26] The industry logged impressive gains during the war years. Domestic production accelerated from 3.8 million barrels per day in 1941 to almost 4.7 million barrels per day by 1945.[27] Refineries also grew in capacity and were modernized, with the average size of refining plants rising from 9,000 barrels per day in 1939 to 13,400 barrels per day by 1945. American oil was a crucial input to more than 500 products central to the Allied war effort, many of which few civilians were even aware of: synthetic rubber, toluene (an octane booster in gasoline fuels, an ingredient in TNT, and an industrial solvent), fuel oil for flame throwers and smoke barrages, range oil for stove and space heating, and medical products. Achieving these gains in output necessitated a tremendous addition to the domestic oil infrastructure along the entire length of the supply chain. Over 17,000 miles of oil pipeline were relaid or reversed in the country during the war.[28]

The war effort facilitated a closeness of partnership between the federal government and the domestic oil industry that would have been unimaginable in peacetime. Oil executives since the days of Rockefeller, who had seen his empire fractured by decree of the federal courts, were wary of government intrusion, even suspicious that federal attempts at regulation might mask a more ambitious agenda, like outright nationalization. That attitude of mistrust did not disappear during the war, though oil executives, however begrudgingly, did acquiesce to strict federal oversight given such extraordinary circumstances. When the OPC was subsumed into the newly formed Petroleum Administration for War, or PAW, in 1942, Harold Ickes continued in his role of overseeing the domestic oil industry as the agency's coordinator. The powers of the federal government during the war to regulate the American oil industry through the PAW were expansive. It reserved the right to fix prices at every stage of production, control production volumes, allocate resources, and supervise the refining, transportation, distribution, and marketing of petroleum. In the words of two historians, "The exigencies of war required a command economy; ideological and material disputes over this fundamental question were suspended for the duration. . . . The inevitable disagreements and debates took place in an arena where settlement, compromise, or concession was possible."[29] The PAW was staffed almost entirely with personnel from the oil companies themselves, which helped ease the transition. Some in Congress grumbled over the arrangement, of course, given the obvious conflicts of interest. Ickes unsurprisingly was more than happy with it, given the unprecedented power it afforded him to manage the affairs of one of the world's most powerful industries.

Less than six months after the United States declared war on Japan in December 1941, the soon-to-be PAW imposed the rationing of domestic consumption of gasoline and other oil products. Ballooning demand, made worse by transportation bottlenecks and inadequate infrastructure, threatened inventories along the Atlantic coast and in the Pacific Northwest. By May, Ickes had put in place a full-blown gasoline-rationing program on the East Coast. Gasoline retailers required that customers present a government-issued coupon at the point of purchase. The system was not popular—"the outcry in the East was deafening," reported one observer.[30] Some of the anger stemmed from the fact that only one sliver of the country had been subjected to rations, which seemed unfair to those along the East Coast who were made to suffer. The problem was not that there was not enough oil to go around—there was. It was just located in the wrong part of the country. It was not easy to get enough trains, railcars, and pipelines to move it from the refining hubs (the Gulf Coast, the Midwest, and California) to the Eastern cities and ports. Ickes, responding to complaints from Eastern politicians about their constituents being the only ones to face gasoline rations, extended the rationing to the entire domestic market in December 1942.[31] The goal of the policy shift was to pool the nation's gasoline supply as much as possible so that areas of greatest shortage could draw from those of greater relative abundance.

Ickes and his colleagues feared that the country's oil reserves were already in permanent decline. This conclusion was not outlandish on its face, given recent trends. New reserves were not being booked as easily as they had in the past, despite the number of wildcat wells drilled jumping from 2,224 in 1937 to 3,045 in 1942.[32] What had been a tight but manageable oil market in 1942 appeared increasingly grim entering 1943, even as the buildup of American troops in Europe and Asia was accelerating. For Ickes, it was time to warn the American people about the trouble in the oil market that lay ahead. He took to the airwaves in February 1943 with a sober message for his national audience:

> If it were possible for me to tell you just how much oil will be needed to carry out the tremendous Allied offensives planned for 1943, you would readily understand my failure to be optimistic as to [sic] future as far as concerns civilian supplies. However, I can tell you that doubling or trebling the amount of oil shipped across the seas will result in a great deficiency in civilian oil stocks. And this gloomy picture of the oil supply situation holds true for the West Coast as it does for the East Coast and the Mid-West.[33]

The best one could hope for was that this was a short-term logistical problem that more pipelines or tankers could fix, and not the beginning of the end of American oil supplies. Other higher-ups in the Roosevelt war apparatus were no more optimistic. PAW's director of reserves described the national energy situation in bleak terms: "The law of diminishing returns is becoming operative. . . . As new oil fields are not being formed and as the number is ultimately finite, the time will come sooner or later when the supply is exhausted." Oil production in the United States would never again be what it had been in the good old days. "The bonanza days of oil discovery, for the most part, belong to history."[34]

Ickes had been warning Roosevelt of the declining ratio of proven U.S. reserves to annual production for several years now. By 1943, the day of reckoning appeared to be at hand. An article from a July 1943 issue of *Time* magazine ("Out of Gas?") put the issue in stark terms characteristic of the time:

> Will the U.S. run out of oil in a few years? Yes, said Navy Secretary Frank Knox last week before the House Naval Affairs Committee. He predicted a serious shortage of crude oil within a year. Within 14 years the U.S. will be a "have-not" nation in oil, he said, when the last drop of its oil fields is drained (if the present rate continues). The same day, Assistant Deputy Petroleum Administrator Robert E. Allen warned that the U.S. is threatened in two years unless "miraculously prompt" discovery of new fields offsets declining production. The U.S. must bring in 20,000 new wells a year, said he, to maintain present production.[35]

It was true that drilling activity had dropped sharply after 1941, with exploratory wells taking the hardest hit. This decline in exploration had reduced the rate of growth of domestic crude oil reserves to the point that 1943 production exceeded additions to new reserves by a small margin. The PAW feared that should this supply trend continue, with wartime consumption multiplying, the Allies could indeed find themselves in serious trouble.[36]

A New "Era of Scarcity and Higher Prices"

The end of American oil, now a threat making headlines, recalled the dreary premonitions of a handful of prophets of shortage who had been crying in the wilderness just half a decade prior. These predictions had often carried the implication that the draining of U.S. oilfields, besides sending oil prices permanently higher, would also spell trouble for American military supremacy. But, coming as they did at a time when the country was just getting over a glutted oil market, few people or newspapers paid them much attention.

There had been shortage prophecies in the prior decade, some widely cited, but no palpable sense of national worry in the popular press. At its national meeting in San Francisco in 1935, the American Chemical Society had pro-voked a "violent controversy" over the future of oil by predicting a new "era of scarcity and higher prices" due to arrive between 1940 and 1943. Now events appeared to be bearing out their gloomy vision, refocusing national attention on the details of their prophetic warning. The imminent permanent scarcity of crude oil, the scientists had warned Americans, would translate into a perma-nent rise in gasoline prices—"when prices go up, they will stay up." This era of "scarcity and higher prices in petroleum products" would arrive "long before the total exhaustion of American oil fields," so the long-debated issue of how much oil was still in the ground was ultimately a moot point (though it claimed that total domestic supply was "fairly accurately known" to be between 10 and 12 billion barrels).

In the eyes of the scientists behind the report, the consequences for the United States of this imminent and long-lasting tightening of oil markets would not be pleasant. Some of these predictions turned out to be impressively pre-scient, others woefully wrongheaded. The American Navy and Air Force, the Society argued, would be obliged to assume the "new and onerous duties" of protecting trade routes around major oil exporters. The auto industry would be forced to undergo a "radical shift in auto styles toward light, cheap cars" able to travel further on a gallon of gasoline. Substitute products could not be brought to market quickly enough to avert the shortage. Even the two substitutes they saw as the most promising—oil shale and petroleum extracted from coal—could not help the nation in an emergency, let alone in a war, the outcome of which would be finalized long before such products could be brought to market at scale. And forget about ethanol, the much-vaunted substitute for gasoline that had gained an increasing number of supporters since the 1920s. It would not be viable until downstream prices were five times what they had been in recent years. But the American Chemical Society scientists were dead on about one consequence of rising U.S. oil consumption: Soon, the only way to avert a shortage of oil in the United States would be by importing a lot of what the country needed. South America appeared the most likely source of the imports, though Russia and Persia were possibilities as well.[37] All in all, the American Chemical Society's 1935 warning had envisaged a new oil era about to overtake the country. Endlessly rising prices, imports, and a scramble for substitute energy sources would mark the history of U.S. energy from the 1940s onward.

Oil industry leaders had protested such predictions adamantly. "Fears of an imminent gasoline shortage in the United States, with skyrocketing prices, are . . . exaggerated," they protested. While they conceded that the country would "undoubtedly be faced with a shortage of natural petroleum at some time," it would be "a great deal further in the future than five years." In their opinion, "known" domestic reserves of 13.25 billion barrels would satisfy national demand for at least thirteen years, even without any new discoveries. The 600 million barrels of crude discovered in 1934 alone, if representative of future reserve growth, were capable of satisfying demand well into the future without resorting to additional imports. Moreover, even if prices did rise in the future, higher prices ought not to have any impact on public consumption. They also noted that the Society was wrong in claiming that a decrease in supply from one year to the next inevitably meant that prices were fated to rise. That was far too simplistic a description of the way the market worked. They cited recent research by "government statisticians" that had "not discovered any link between gasoline consumption and price, or between petroleum supply and price."[38]

Facing soaring wartime consumption, industry representatives continued to reject the notion that the country was running out of oil. American reserves were far vaster than the American Chemical Society and other skeptics believed, they argued. One of the chief scientists of Standard Oil Development Company, the company's technical subsidiary, spoke out against rumors of an impending supply shortage facing the country, contending that it would be at least 300 more years "until oil runs out of this earth." His "temperately optimistic" view assumed that the current rates of production would last indefinitely, new discoveries would come about "where they reasonably can be expected," and that oil companies like Standard would be able drill for oil without any meddling from Washington. According to his calculation, the United States had been blessed with reserves containing more than 100 billion barrels of oil. Nearly half had already been used, he calculated. But the rest of the world would likely yield some 500 billion barrels more before all the oil was gone.[39] Neither figure, in his mind, was any cause for worry.

The real problem that had oilmen grumbling as the war raged on was that Washington was not giving them the economic incentives and vital materials, like steel, they needed to keep oil production growing. It was true that the PAW had put in place aggressive controls on the price of crude oil. The OPA had put in place an informal price ceiling as early as June 1941. Each oil-producing state, whether California, Oklahoma, or elsewhere, was assigned its own price level, intended to allow a "reasonable margin of profit" depending on the local costs of production. It did not take long, though, for the OPA and later the PAW to see that these price controls stood in the way of stimulating the enormous new production that the war effort required. If the OPA and the PAW would allow market forces to operate more fully, they argued, the supply problem would solve itself. But the Office of Economic Stabilization, to which the PAW appealed routinely in 1943 for price increases, was loath to let crude prices rise. All said, the OPA would only allow oil prices to rise twice during all of World War II. Shortages of steel were chronic throughout the war, and the oil industry often felt it was asked to do the impossible in expanding drilling and pipeline infrastructure while being allotted insufficient materials. The best evidence of the glaring inefficiencies in the upstream supply chain, under the management of the PAW, was the startling decline in the number of wells drilled over the course of the war, which fell from 32,053 in 1941 to just 19,431 by 1943.[40]

The controversy over the government's fixed price for oil was the subject of a hearing before the House Subcommittee on Interstate and Foreign Commerce in 1943. Robert Fall, speaking for the oil industry, attributed the recent decline in production solely to the artificially low prices the government had fixed for oil products. He argued that plenty of oil was to be found, but that the current crude price ceiling, which had locked in 1937 oil prices, was inadequate to stimulate new discoveries. Without the OPA raising the price per barrel by at least 25 percent, domestic production would dwindle. Walter Hallanan, president of the Plymouth Oil Company, issued a similar warning: "It is now a great national and international problem which seriously threatens the effective prosecution of the war and holds grave consequences for our domestic economy." Only a price increase, in his view, could do the trick. As for the OPA's assertion that such a price increase was "unnecessary and inflationary," it was "just plain bunk," remarked Hallanan. "What would it profit to hold the line against inflation and perhaps lose the war as a result, or at least prolong it at a terrible cost in the blood of American youth?"[41] Despite the impassioned testimony, the OPA was not persuaded. While it did allow the price of some products to increase, it chose instead to attempt to stimulate production through granting subsidies for exploratory drilling and rewarding any attempts to extract oil from older wells, a process known in the industry as stripping. The subsidy program paid out $64.9 million in 1944 and 1945 to 308 operating companies. Despite the doubters, it did help give production a lift. One estimate attributed the production of 177 million barrels of crude as a direct result of subsidization.[42]

The Nightmarish Thought of Imported Oil

In one essential respect, the national debate over oil depletion during World War II began to differ fundamentally from previous eras. A growing consensus among policymakers, experts, and industry leaders began to emerge amid the strain of America's effort to fuel the Allied war effort. For them, what lay ahead for the U.S. oil economy was simple: domestic reserves were increasingly unable to meet the country's demands at the prevailing price. They might even be running out for good. Either way, the United States would have to import oil, in increasing quantities. Foreign oil was the only and unavoidable answer, barring a drop in U.S. demand, which did not look likely. Something had to give.

In this emerging consensus of U.S. energy thinkers, domestic exhaustion did not mean the end of the global oil trade; but it did mean that locus of production would shift from the United States to those regions of the world that had not been pumped as long and as hard as the home reserves had. As acclaimed American geologist Everett DeGolyer summed it up in 1943, "Our low rate of discovery seems not to be a result of slackening of effort so much as due to the low quality of prospects, which indicates an exhaustion of prospects discoverable by the present methods of utilizing known techniques." Of course, it was foolish to see the "known techniques" at the time as the outer limits of what technology would ever allow, but some analysts did feel the country's oil production may not be going much higher in the decades ahead. In their view, "the problem was believed to lie in the nature of the resource itself—the absolute quantity of petroleum available," which would soon be exhausted domestically.[43]

The thought of the country's becoming a net oil-importing nation was an unwelcome thought to strategists in Washington, who were accustomed to planning under the presumption that the country would always have all the energy it needed in a contingency. It was alarming news that there was "a real possibility," as one oil executive put it, "that the day is not far off when the United States, long known as the world's greatest exporter of oil and oil products, may have to resort to the importing of that commodity to meet her needs."[44] Government leaders—Ickes and PAW, Roosevelt, Congress—found the idea distasteful. As they saw it, the country's ability to provide almost enough oil to meet its domestic and military needs in both World War I and World War II was a strategic blessing that would be hard to do without. They had never had to call on other nations to meet their own energy needs with any sort of regularity. Yes, there had been forays into Mexico, Venezuela, and other oil-producing states prior to the 1940s. But those ventures were for corporate profit and international prominence, not for national need. Politicians long favored their home-team companies acquiring overseas reserves—that could only increase the country's economic and political standing, if done prudently. But actually depending on imports from foreign countries to satisfy the country's demand for a vital strategic commodity was another thing entirely. For the White House, it could mean a painful loss of leverage in diplomatic relationships with oil-producing states. It was anathema to the ideal of self-reliance—the ability to fend for oneself—an ancient American virtue. The abundance of American oil was also a nice carrot to offer U.S. allies, who could depend on a reliable flow of oil to come directly from the United States in a time of war.

For those in the oil patch, the fact that the country had almost always produced enough to spare spoke to the entrepreneurialism of its businessmen, the technical savvy of its engineers, and the God-given bounty of the national landscape. These were all points of pride for American oilmen. On top of it all, business was just plain easier for U.S. oil companies, especially the small-scale independents, when they did not have to look any further than their own backyard to find plenty of giant fields. The largest American oil companies, such as the former Standard firms, were vertically integrated internationally. Even if the most promising new fields were found overseas, they would be fine, given their adeptness overseas across the globe. If domestic reserves were indeed nearing exhaustion, as many believed, it was the small firms that would suffer the most. If the international companies were to flood the American market with cheap foreign oil, the small domestic players knew they might not survive.

The new accepted wisdom—that American oil reserves could not satisfy the war effort much longer and would soon become extinct if demand growth continued unabated—prompted government leaders to reevaluate their oil strategy. In a country of dwindling reserves, the wisest course of action appeared to many to be to conserve domestic supplies while buying up concessions and producing oil abroad as intensively as possible. A letter from Michigan Senator Prentiss Brown to Ickes in 1943 summed up this emergent line of reasoning:

> In your two letters you have dwelt at length on the necessity for increased domestic crude production, yet I believe your most recent approach to the supply problem, namely, through increased use of foreign reserves, constitutes the real answer. Since it is true that the reserves of this country are not inexhaustible, it seems only logical that our military and civilian requirements should be satisfied to the greatest extent possible through supplies available in other countries. If "oil is ammunition," according to the popular slogan of the oil industry, let's conserve our ammunition to the greatest possible degree. . . . Foreign crude oil and residual oils are available during the war emergency and should be used to the maximum possible extent in order to conserve our domestic resources for future years.[45]

The United States simply could not continue to produce as it had up to 1943, went the argument, without becoming a "'have-not' nation in oil," as Navy

Secretary Frank Knox put it. American resources needed to be conserved and those abroad needed to be developed. The United States should save all it could, lest it one day find itself depleted and totally dependent on access to foreign oil.

At a Senate investigation of the National Defense Program in October 1943, a committee of five senators called on federal officials to put in place whatever restrictions were necessary for the country to curtail the pumping of its own oilfields, and opt instead to take crude from other countries. "The United States must rapidly move to preserve so far as practicable its rapidly diminishing reserves," they warned, at the risk of Washington being "unable to face the world on an equality at the end of the next decade." One senator from Georgia lamented that "up to now, we have been depleting our petroleum stocks at a ruinous rate, supplying not only our own forces but those of our allies. It is now time to utilize the petroleum deposits of other parts of the world. Otherwise, the end of the war will find our own deposits practically exhausted." Selfish Britain was especially at fault, they felt. It was said to have oil reserves "at least equal" to the United States, and yet "only 8 percent of the oil consumption of the United States is coming from British sources, while 80 percent is coming from United States reserves." If the trend continued, America may have to go hat-in-hand to foreign nations for oil "before another generation comes on stage," a "mendicant for petroleum at the council tables of the world," unable to supply its own needs.[46] Something needed to be done.

"The Center of Gravity of World Oil Production Is Shifting"

The oil shortage fears of 1943 took the attention of the U.S. government to a nation that Roosevelt had dismissed as "a little far afield" for political involvement of any kind only two years earlier: King Ibn-Saud's Arabia.[47] Two American oil companies, SoCal and Texaco, had struck oil there in commercially viable quantities in 1938. They had pressed their home government early on to support their venture, sensing both Nazi and British designs on the area, but to no avail. As the war advanced, however, Ickes asked Roosevelt to reconsider helping to secure American oil operations there. The federal government had a sense of what an enormous amount of oil existed in Saudi Arabia thanks to Everett DeGolyer, the preeminent petroleum geologist of the day and a deputy director to Ickes's PAW. Ickes had sent the geologist to the Persian Gulf in 1943 to assess the region's oil resources.

DeGolyer came home with an astonishing verdict that foreshadowed a dramatic change in the geography of the world oil market. He estimated the proved resources in the Middle East at 16 billion barrels, and the probable reserves between 25 and 27 billion barrels. The estimate exceeded Washington's expectations by a mile—as well as American reserve estimates by several orders of magnitude. In reality, this initial estimate of the total resource base would prove to be far too low. Current estimates of proven reserves in the Persian Gulf area come in at over 650 billion barrels, not including probable and possible reserves. The note that DeGolyer wrote in 1943 about the future of Middle Eastern oil supplies would spark a prophetic reassessment of the global oil order by his colleague, John Murrell, reprinted for the public in November 1945: "The center of gravity of world oil production is shifting from the Gulf-Caribbean area to the Middle East—to the Persian Gulf area—and is likely to shift until it is firmly established in that area." Despite the "highly speculative" nature of these early resource assessments, they arrived at the "inescapable" conclusion that "reserves of great magnitude remain[ed] to be discovered" in the region.[48]

Anxiety within the PAW in late 1943 over an impending oil shortage in the United States led Ickes to conclude that the best course of action might be for the U.S. government to buy the Saudi oilfields outright. The Army Navy Petroleum Board told the American public in 1943 that its projections for the following year were not looking good. "A serious shortage of oil," it concluded, could "threaten military operations." The possibility of the Allied offensive stopping in its tracks because the United States failed to fulfill its pledge to keep the oil flowing was unthinkable. Ickes decided to do something. He contacted the presidents of Texaco and SoCal with an offer: sell the federal government your entire Saudi oil operations. Washington was going into the oil business. The government oil company operating abroad would be known as the Petroleum Reserves Corporation, or PRC, Ickes told them. And it was also exploring a similar deal with Gulf for the Kuwait concession. Facing resistance, Ickes lowered his proposal for 100 percent ownership to a simple majority ownership of 51 percent. Finally, the deal was struck: the government would purchase one-third of Casoc (also known as California Arabian Standard Oil Company—Chevron and Texaco's jointly owned Saudi Arabian concession) for $40 million, including the right to buy up to 51 percent of Casoc's production in peacetime and 100 percent in wartime.[49]

When news of the acquisition spread, the rest of the oil industry exploded with protest. But Ickes, who was to be the first president of the PRC, defended the move on the grounds that the United States was one of the few great powers that was not working with its domestic oil industry to secure overseas drilling acreage. Ickes eventually abandoned the Casoc deal in the final months of 1943, bowing to intense pressure from independent oil producers and most of Congress. But he had not given up on the idea of the PRC getting involved in the oil trade.

With Casoc scuttled, Ickes devised a new plan: a $120-million pipeline across the Arabian Peninsula, sponsored by the PRC, which would bring Persian Gulf oil to the Mediterranean for export to the Atlantic. A Senate subcommittee report released in February 1944 summed up the energy crisis the country was facing. "The depletion of our petroleum resources does not present the possibility of an immediate catastrophe," the senators wrote, "because fortunately we have, beyond doubt, sufficient petroleum to win this war." But whether future supplies would be there was less certain. It was doubtful that the country could "oil another war." The future of "our national safety and a continuation of our industrial progress hung in the balance," they argued. "The security of the United States today is dependent upon adequate supplies of petroleum." Congress needed to do something. The committee proposed the formation of the PRC as a means of ensuring that the country would have ready access to the crude oil it needed after the war, both for military and economic reasons.[50]

Top commanders in the U.S. armed forces were outspoken in advocating for the PRC on national security grounds. Secretary of the Navy Frank Knox, outspoken on oil issues, warned that the United States was depleting its oil resources "very rapidly." It was "selling its oil as if our supply was inexhaustible, but it isn't." The "dangerous rate" of extraction, he added, came as the result of letting "a lot of selfish oil companies" derail the implementation of a sound national energy policy. Were another war to come, a lack of oil could prove disastrous. He reminded Americans that 90 percent of "war oil" came from North and South America. Unless the United States "began to develop new sources for war needs, the country might find itself becoming a have-not nation in petroleum."[51] Even Roosevelt, who rarely spoke publicly about the oil situation, told reporters at a March 1944 press conference "his concern about supplies." The president expressed concern about "the nation's oil resources for not only the next five years but also for the next fifty."[52]

Despite military support and White House sympathy, congressional and industry indignation at the nationalization plan eventually doomed the PRC's playing a role in the U.S. oil trade. After months of trying to persuade Congress and the American public that nationalization was the right move, Ickes let the plan die out slowly and quietly in 1944. It would be the first and last time the U.S. government would attempt to get into the oil business directly. The oil industry had few kind words for the plan even after it had been scuttled. For American oil executives, the spate of "alarmist" reports by the PAW and military in 1943 were nothing more than Washington propaganda aimed at taking over their livelihood. In their eyes, U.S. officials had tried to stir up the perception among the public that oil supplies were running out in order to scare them into acquiescing to their bid for control. As the *Oil & Gas Journal* put it in March 1944:

It is apparent that millions of consumers are convinced that a permanent shortage exists in this country, and it is on the fears that arise
from this belief that sponsors of the governmental participation in
oil operations here and abroad are largely backing their case. The average consumer still does not realize that petroleum rationing has
been brought about entirely by the war's restrictions in operation.
There is danger that he will support any program which he is led to
believe will assure adequate petroleum supplies in the future, regardless of economic and political complications to him and to future
organizations.[53]

No matter the shortage that may or may not be looming, critics of the plan saw
it as a bridge too far in the direction of state-run capitalism. In the words of one
industry lobbying group, the PRC's proposed trans-Arabian pipeline, which
was to run from the Persian Gulf to the Eastern Mediterranean, was a "move
toward fascism." It reiterated the view that the "real causes" for the decline in
oil discoveries and capacity increases in the United States during the war were
artificially low prices, administered by Washington, and temporary shortages
of labor and materials, not a lack of oil in the ground.[54]

When the war finally ended, coming to terms with the new playing field in
the global oil market became a full-time preoccupation for the industry. The
days of endless worrying over the duration of the war—and whether Allied oil
supplies could be maintained—were finally over. There was little question that
U.S. oil companies had been remarkably capable in meeting the strenuous demands that World War II had placed on them. PAW had gotten better and
better at its job, allocating and delivering material resources to the oil industry
more efficiently. Domestic oil production had climbed from 3.8 million barrels
per day in 1941 to 4.7 barrels per day in 1945. The jump had required slightly
exceeding maximum efficient recovery levels (the rate of production above
which the ultimate recoverable reserves in a given field begins to drop) by mid-
1944.[55] Victory had eased the demand equation. Gasoline rationing was lifted
within 24 hours of Japan's surrender, in almost celebratory fashion. The wartime oil bureaucracy was all but completely dismantled by the close of 1946, as
the OPA finally removed the price controls it had imposed during the war and
eliminated the subsidy it had paid for the operation of stripper wells.

"The Inexorable Demands of Self-Preservation"

The resolution of World War II did not end the global contest for control of
prized oil resources; it only reshaped the playing field on which the game was

played and saw the rise of new energy adversaries. The postwar world appeared to American policymakers to be splitting in two—one part American, the other part Soviet—and the world's oil supply, they believed, would likely be divided along with it. The schism heightened the urgency of the rush for securing oil supplies. American military strategists worried that America's allies might not have enough of the critical commodity should another war begin. It was a matter of necessity that the United States "organize itself for a protracted confrontation with the Soviet Union," and oil had a "central place in the strategy for security" in this nascent rivalry.[56] In the words of a Senate report in February 1947, Washington needed to put in place an oil policy well adapted to "this troubled world, which has not yet learned how to avoid war.... The oil policy of this nation while at peace must nevertheless be governed by the inexorable demands of self-preservation." Rather than "await with hope the discovery of sufficient petroleum within our boundaries," the far better course was to provide government incentives for exploring for oil domestically while trying to manufacture "synthetic liquid fuels" beyond oil.[57]

The question of whether American oil supplies were sufficient for the task of defending U.S. interests took on new urgency. "The Navy cannot err on the side of optimism" in estimated domestic reserves, said Navy Secretary James Forrestal, whose name would eventually come to grace the façade of the U.S. Department of Energy headquarters, in 1945. "The prestige and hence the influence of the United States is in part related to the wealth of the Government and its nationals in terms of oil resources, foreign as well as domestic. The active expansion of such holdings is very much to be desired."[58] Forrestal expressed his worry to Secretary of State James Byrnes the following year:

> I am of the opinion that [Ickes] is right about the limitations of American oil reserves—in this I am influenced a great deal by the engineer that I used in private business, E. L. DeGolyer. If we ever got into another World War it is quite possible that we would not have access to reserves in the Middle East but in the meantime the use of those reserves would prevent the depletion of our own, a depletion which may be serious in fifteen years.[59]

Data from the domestic upstream reinforced the tense climate. The annual net increase in proved reserves between 1935 and 1939 had averaged 1.3 billion barrels, a comfortable buffer. During the following six years, however, despite new discoveries averaging 1.9 billion barrels annually, the additional consumption demands of the war resulted in a net annual increase in booked reserves of only 0.4 billion barrels.

Not every American oil analyst thought that production was doomed to struggle in the years ahead. A U.S. Tariff Commission report in 1946 acknowledged that there were two schools of thought regarding the future of American oil. While "some experts believed the United States was doomed to rely increasingly on imports," others argued that "the decline in discoveries during the last few years has been due largely to wartime conditions and that now that the war is over the rate of discovery will rise again." The question was whether the disappointing rate of discoveries during the war was the beginning of a new era of increased import dependence, or less hauntingly, an artifact of a wartime regulatory apparatus that did not do enough to promote domestic exploration, which would prove ephemeral. "Pessimistic forecasts regarding the future supply of petroleum in the United States, which have been put forth at intervals for a long time past, have often been based too largely on the current estimates of proved reserves" without considering future discoveries, they observed. But the increase in reserves in recent years had been "much less marked in the last few years than before," which might portend a turning point. Should the ratio of new discoveries to current production continue to decline, "the time would soon come" when the country would produce more oil than it could discover, and a permanent shortage would set in.[60]

"Much More Difficult and Far More Costly"

A string of statements by top U.S. officials in 1946 and 1947 underscored their concern about a looming lack of oil. For them, rising oil imports presaged a new era of climbing oil prices as the quality of global reserves waned and the costs of transporting the crude added to the price of oil in the United States. In August 1946, top energy analysts from the State Department and members of the Army-Navy Petroleum Board took to the airwaves to warn Americans that they faced an oil shortage in twenty years unless Congress passed the stalemated Anglo-American Oil Agreement (whether "shortage" in this context meant an uptick in imports or an absolute lack of oil is hard to tell). The intention of the proposed legislation was to develop a joint British-American decision-making body that would "balance discordant supply and demand, to manage surplus, and to bring order and stability to a market laden with oversupply."[61] The strategy that London and Washington used to garner public support for the measure was to highlight the uncertain future of their own domestic oil resources. The chief of the State Department's Petroleum Division warned, "I'm no prophet and I don't propose to get into a statistical battle. But I think it's safe to say that by 20 years from now, we shall have to import close to half the oil we consume—unless some large new oil fields are discovered, and we can't count on that."[62] The broadcast, which "created a sensation," led the *National Petroleum News* to reassure the public that it "does not believe that any disastrous shortage of oil supplies will be on us tomorrow." It cheerfully recited the ways in which producers were finding ways to extract more oil from unconventional sources and refiners might one day cheaply process oil shale and coal into liquid fuel.[63]

Secretary Julius A. Krug, Ickes's successor as President Harry Truman's newly appointed secretary of the interior, took to the airwaves in the summer of 1946 with bad news for anyone who cared about cheap oil. Prices were going up, he argued, perhaps indefinitely into the future, as American oilfields fell into decline. "Expert opinion as to the quantity of oil we may hope to find through future exploration indicates that we have reached the halfway mark in oil discovery in the United States," he announced. The future would not be as rosy as the past: "Finding the second half will take much longer, be much more difficult, and far more costly." The era of American dominance in oil was coming to a close; its current rate of production was no longer sustainable:

> Ultimate oil resources of the United States, according to the experts, will amount to about one-sixth of the total oil resources of the world, and we are producing and using about two-thirds of the worlds' oil. This means that oil will be scarce in the United States long before there is a comparable scarcity in other important producing areas. . . . I believe the industry must accommodate its thinking to the need for substantial petroleum imports which can increase our security without being any threat whatever to a healthy domestic industry.[64]

Krug saw the writing on the wall: Oil was simply cheaper to produce in places like Arabia and Persia than it was in most U.S. fields. He knew, rightly, that production would flow to the place it could be done the most cheaply. The only logical conclusion was that the United States must come to terms with the new reality that foreign oil supplies would fill a larger, perhaps even dominant, share of national consumption before long. But his calculation that American oilfields had already entered terminal decline, and that with ballooning imports would inevitably come higher and higher oil prices going forward, was off the mark.

Not "as If We Might Run Short of Oil but as If We Had"

Apart from the turmoil about if and when new supplies would be found, the other side of the oil market was undergoing dramatic change: Demand was soaring. The postwar boom in American population growth, multiplied by a people getting richer, drove the nation's appetite for oil to levels that scarcely would have been imaginable even half a century prior. The country's consumption of gasoline and diesel doubled between 1939 and 1950, hitting 2.7 million barrels a day. For distillates, it almost tripled. Oil consumption ballooned by 32 percent in the five years after the armistice alone.[65] The wheels of the country were turning, with vast numbers of cars, trucks, and buses hitting the road. Manufacturers could scarcely make them fast enough to fill orders. The 26 million cars on American highways in 1945 would turn into roughly 40 million by 1950.[66] And transportation was not the only sector posting gains. The country bought 40 percent more space heaters and oil burners in 1950 than it did the previous year.[67]

Facing this spike in demand, the industry struggled to develop the infrastructure it needed to transform a booming wartime energy economy into prosperous peacetime one. Refineries had to be reconfigured to produce the kinds of products that everyday consumers wanted, like heating oil, versus the kinds of fuels that military operations called for but civilians had little use for, like aviation fuel for fighter planes. Making the switch was time-consuming and expensive.[68] Steel, vital for expanding the country's overwhelmed oil infrastructure, was still scarce, making transportation bottlenecks a problem. Oil companies launched a multi-billion-dollar infrastructure investment effort in 1947 to try to solve the problem. But it was too little, too late. Although there was still plenty of oil in the ground, it could not be brought fast enough to where it needed to go before a double whammy in 1947—the arrival of an ever-more-frenetic driving season followed by a brutally cold winter.

Beginning that summer, shortages began to appear in certain parts of the country, affecting both regular consumers and U.S. government buyers. Standard Oil of Indiana had to resort to rationing the gasoline it sent to service stations in the Midwest.[69] The director of the Department of the Interior's oil and gas division, Max Ball, forecast a worsening shortage that winter if the weather turned especially cold and a maritime strike underway lagged on.[70] The Army-Navy Petroleum Board announced in July it was already short of the aviation gasoline it needed to meet the military's immediate needs, causing the Department of Commerce to ban gasoline exports until government inventories rose. The Navy was ultimately forced to import 3.4 million barrels from the Persian

Gulf—a desperate measure, in its eyes.[71] In a Cold War environment, the prospect of the military not having the petroleum products it needed to protect and defend made the situation an uncomfortable one in Washington.

As national discontent over rising prices grew, accusations in Washington flew over who bore responsibility for them. Congress held over twenty hearings about oil prices and shortages in 1947 and 1948. "The oil industry's supply and demand problem getting bigger and blacker headlines in the nation's newspapers and more space in the Congressional record this week," said the *National Petroleum News*, not wholly approvingly. "The industry itself still feels that the situation is basically unchanged," with transportation bottlenecks in the Midwest keeping oil away from customers on the coasts.[72] Politicians railed against American oil companies for continuing to export crude despite the shortfalls at home. In June 1947, The Commerce Department, denounced the fact that nearly 35 million barrels of oil and oil products had been shipped to Russia, Europe, and Canada that year—the same amount that had been shipped during 1946, when there was no shortage. One senator accused the oil companies of "feeding the war machine of Russia while Russia refuses to cooperate in the world program."

Oil companies responded to the indictments with barbs of their own. They lambasted steelmakers for failing to give them enough supplies to build out their infrastructure so that they could get the oil where it needed to go. "Communist Russia alone, in the last 20 months, has received 50,000 tons of pipe, casing, and other tubular goods," even while the American oil industry lacked those basic goods, cried the National Petroleum Council.[73] Executives criticized Congress for making the oil industry a scapegoat for market conditions that were unsurprisingly tight, given the huge surge in consumption, and thereby turning a stressful few months into a full-blown public panic. The industry appealed for calm, denying that these seasonal shortages would turn into any bigger a problem. "You hear the statement that there is an immediate danger of our running out of oil," said Harold B. Fell, the executive vice president of the Independent Petroleum Associations of America. "The fact is, there is no shortage of oil. Discoveries of new reserves are continuing. Known proved reserves are 22,000,000,000 barrels—an all-time peak. There is no danger of an oil shortage. Competent geologists have testified that there is at least as much to be found."[74] Some oil operators lay the blame for the scare squarely on their more powerful brethren. A group of Chicago oil dealers, for instance, claimed the shortage was nothing but a "trumped up device of the major petroleum producers to drive independent distributors out of the business," according to the *Chicago Tribune*. They told Congress that the shortage rumors were all a hoax, which the majors had propagated for their own benefit.

But it was apparently all too easy to misinterpret the situation for something more dire: that the United States was falling prey to a chronic lack of oil to service its basic needs. The trouble in the market, which prices and public pronouncements appeared to be signaling, "gave the Cold War oil scare support of a most sensational and urgent, if circumstantial, nature. It not only appeared to the man on the street as if we *might* run short of oil but as if we *had*."[75] Rising oil prices would have done little to allay such concerns. By 1948, they were more than double where they had been in 1945, when Washington had been capping them. Gasoline prices rose nearly 25 percent between July 1946 and December 1947, when they hit $0.18 a gallon.[76]

As acrimonious as the rhetoric was in the summer of 1947, it only got worse as the weather cooled. The country's oil fields were pumping more than ever before, it was true. But transportation bottlenecks—not just in the United States, but around the world—were still a problem for the oil industry. Heating oil inventories were problematically low. Representative John W. Heselton of Massachusetts, striking an alarming note, warned the nation that the country would completely run out of fuel oil by February 20, 1948. In October, oil companies were forced to begin to ration their inventories, refusing to service new customers and carefully guarding whatever they had on hand.[77] For many Americans, including U.S. officials, it became increasingly hard to tell whether this was just a temporary shortage stemming from a lack of transportation capacity or if the nation was simply beginning to feel the hunger pangs of a new era of oil scarcity. It was the former, of course, but the fog of war was starting to settle in over the country's debate about the oil market. The *New York Times* explained the situation correctly in November, "Today there is enough oil in proved fields throughout the world—50 billion barrels is a conservative figure—to meet all our needs. But there are not enough ships, pipelines, tank cars, barges and refineries to get it where, and in the form, it is needed."[78] Oil executives, as usual, had no patience for the "exaggerations" of media reports of the impending "gasoline famine." They attributed an apparently widespread belief across the country that service stations around the country may run completely dry at any moment to these misleading or simply misinformed news stories.[79]

"It's Finally Happened"

But there were plenty of others—including members of Truman's cabinet—who did plenty to fan the fires of public panic over a looming oil shortage, which might be permanent. Secretary Krug, head of the Department of the Interior, had bleak words about the energy catastrophe that was about to hit the country. "Today's oil shortage is not just temporary—it's for keeps, and it's going to get worse," reported the *Christian Science Monitor* in summarizing Krug's statements:

> It's finally happened—the thing geologists forecast for a quarter century and which the petroleum lobby said wouldn't occur for a hundred years. America's demand for oil finally has outdistanced supply. . . . Ever since the Indians discovered the first iridescent oil slick on the Pennsylvania marshes, it has been assumed the petroleum supplies would last indefinitely. In fact, American production is holding up, though its wells are being pumped unmercifully and its back is being flogged like a laboring horse. But one thing is moving faster—demand. In brief, what has happened in the cold winter of 1947-48 is that the United States has come to the end of an era. . . . The time has come when we cannot depend on natural petroleum alone to meet our needs, he said: we must start converting natural gas, shale and, above all, coal, into oil. "The dawn of a synthetic-liquid-fuels era in the United States is coming—and coming fast," the Interior Secretary declared.[80]

Krug was unquestionably right in part: U.S. oil consumption would soon run ahead of production. But he seemed to all but ignore the potential for imports to help make up what domestic oilfields could not. The United States was becoming a net importer of crude, which would be a permanent change in its role and status in the oil market. But implying that that change meant that oil "shortages" would be the order of the day from there on out, or that "natural petroleum alone" could not meet the country's needs, was a half-truth that did little to discourage alarmist media reports. If oil was running dry in the United States, was there any hope for major discoveries further afield? And would foreign oil reserves, which were much more uncertain, ever be able to provide the same abundance of energy that the country had known over the half-century prior?

The country braced itself for the winter. Truman ordered the temperature in all government office buildings and residences to be lowered to 68 degrees to save heating oil and placed a 40-mile-an-hour speed limit on government vehicles to save gasoline.[81] Forty million gallons of oil were released from the Navy's reserves to supply New England in case of an emergency.[82] The oil industry was able to secure additional tankers to make the Gulf Coast–East Coast supply run, which eased the situation that winter. Refiners on the East Coast churned out heating oil at the expense of gasoline. But even with domestic production at an all-time high of 5.3 million barrels per day in 1947, the country's demand for crude—14 percent higher that year than at the 1945 wartime high—exceeded what the country could produce.[83]

The year 1947 marked an important first for the country's oil economy, which in time would radically change the way Americans viewed oil as an energy source. The year also marked a goodbye to the country's status as a net exporter of oil and oil products that persists to this day. Never again has the United States sent out more petroleum than it received from other countries. The country's days as the world's greatest oil supplier were over. Imports increased by an estimated 13.2 percent between 1947 and 1948, reaching an average of roughly 500,000 barrels per day. It would be an estimated 75,000 barrels per day more than exports would amount to over the same period.[84] From then on, the United States would have to rely on other nations for marginal production sufficient to meet its growing appetite.

By 1948, almost every voice on record expressed skepticism that the so-called oil shortage would ever be resolved. The term "oil shortage" appears to have referred to the fear—which would ultimately be borne out by history—that domestic oilfields would ever again produce enough oil to meet the country's escalating demands, at least in their lifetimes. Some observers began to conclude that the industry's consistently upbeat projections about the future of oil were unfounded:

> It now appears to be only a matter of time until big names in the oil industry begin to admit publicly that the United States can no longer hope to remain self-sufficient in petroleum supplies through the discovery of new oil fields and better ways to get the oil out of them. . . . It begins to appear that in expressing their one-time optimism petroleum executives not only spoke with fingers crossed but were actually hedging their optimism for getting oil from other sources than the fields in the continental United States.[85]

If the United States was having trouble pumping oil fast enough, and it was the largest oil producer in the world, then the future did not appear bright for supplies globally. It was all too easy to extrapolate from growing American imports a wrongheaded view that oil production everywhere was destined to decline soon—and with it a new era of global scarcity about to ensue, making the low prices and abundant production of earlier years a relic of history.

Oil executives, in defending their view that no catastrophic oil shortage was at hand, continued to be optimistic about discoveries yet to be made, both at home and abroad. For them, the kind of "oil shortage" that conjured up images of an absolute dearth of oil in the ground, coupled perhaps with perpetually rising prices, just did not look at all likely for the foreseeable future. In that sense, history would also bear them out, too. Eugene Holman, president of the Standard Oil Company of New Jersey, assured the National Press Club in 1948 that "there is no crisis in oil." He did not share the gloomy view in circulation at the time that there would not be enough oil to go around in the years to come. "No one in this country is going to have to make any important change in his way of life because of a lack of oil products," Holman emphasized. "I say this while fully aware of the great attention being given to supply difficulties, and also knowing that some people accuse me of being an optimist on the supply situation. I think there is a great deal to be optimistic about."[86] That said, many in the oil industry apparently thought that the fuel crunch stemming from transportation bottlenecks could last for three years or more.[87]

Whatever their definition of "oil shortage," there is little question that Americans seemed to view the country's future demands as almost impossibly high for an exhaustible resource to meet, despite oil companies' heroic attempts at "performing miracles of production,"[88] which presaged an era of rising prices. "The $64 question . . . in the oil industry's future," asserted one author, "is 'Whence will the United States get its future oil supplies as its consumption rides steadily higher than its present and future fields can supply?'" Oil companies had gone into the Middle East and Venezuela, he noted, but he had little hope that either of those two places would yield much oil. Today's shortages were just a shadow of the chronic energy problems that awaited a country that was living beyond its means. In the words of the *Christian Science Monitor*: "The fuel oil shortage is no passing seasonal emergency. It will not be over when the winter ends. Next summer gasoline will be in short supply. There is every indication now that it will put a real crimp in pleasure driving. It may even mean rationing. . . . If next winter is a frigid one, there will be another scramble for fuel oil." Unprecedented levels of consumption were more than just uneconomic—they were immoral. It was just not right that, even with all the oil they already had, "Americans want more. They want millions of gallons more than they ever had before."[89]

Still, for an influential group of military leaders and the House Armed Services Committee, the oil question was far more existential than ethical. For them, the oil shortage that was about to strike was the kind of old-fashioned "running out of oil" scenario that had worried Americans several times before. No group was more outspoken in calling for emergency measures to meet a possible oil catastrophe than these officials. In a May 1948 report, the House Armed Services Committee issued the country yet-another red alert about a looming oil emergency facing the United States. "We want to make it as plain as words can—we want to emphasize," declared the Committee, "we want each of you to carry this thought away with you above all others from this report—the nation is in a grave situation in respect to its petroleum." U.S. oil production, and with it the survival of the country, were in jeopardy. The United States had already been drained of "well over half" the total domestic supply of oil. "Only about a 12-year supply of oil remains," the Committee warned. It urged the government to take emergency measures. Among the solutions it proposed were reviving voluntary rationing, setting up military stockpiles, setting aside additional military reserves, funding oil exploration in Alaska, developing the continental shelf, and putting money into research on synthetic fuels.[90]

Other voices in Washington, as well as oil company spokesmen, flatly rejected the new study's reserve estimates. Max Ball, director the Oil and Gas Division of the U.S. Department of the Interior, disputed that the country faced any kind of emergency over oil supplies. Barring a full-blown war, he predicted, "the American public in the years ahead would receive a more abundant supply of better petroleum products at lower relative costs than it has ever known." The Interior Department dismissed the idea that "the petroleum supply would never again be abundant and that world demand would grow too fast for even the industry to satisfy." Yes, it did appear to be true that the United States was going to need to import oil in greater and greater quantities. But that did not mean that oil was truly running out or that prices were destined to keep rising. Ball waxed pedantic: "Such predictions ignore economic history, discount the vigor and enterprise of the petroleum industry, and underrate the driving force of competition." As long as "normal competitive forces are permitted to operate," the oil industry's future looked as bright as ever.[91] It would have been difficult for him to articulate a vision of the future of oil that contrasted more sharply with what Congress had put forward just days before.

The following year, a spokesman for the Independent Petroleum Association of America, L. Dan Jones, testifying before the House Foreign Affairs Committee echoed Ball's assessment: there is no foreseeable exhaustion of oil resources, whether domestic or foreign. "Petroleum is no longer in world shortage," he reassured the legislators. Indeed, as was the case in 1949, world markets are "faced with a surplus of oil which is reflected in the United States by rapidly increasing and unneeded imports of foreign oil." Asked whether he believed the "statements . . . issued some 2 years ago that our known reserves could last only for about 14 years," Jones did not buy it. Dividing known reserves by projected consumption was misleading: it inevitably failed to take into account new reserves that would be added in the future. "Twenty years ago," if you had done the math, "you would have found that we had only 10 or 12 or 15 years' supply and we would have been out 5 years ago." But those figures were misleading, for the simple fact that "we have always found more oil than we have actually used."[92] Such optimism came as a "startling surprise," one member of the chamber observed, as it stood "directly in opposition to the testimony which we have had from the responsible members of the present government."[93]

"The Industry Has Overtaken All Demand"

Even as the debate raged over whether a new era of scarcity was at hand, high oil prices were beginning to bring more oil out of the ground. Prices had soared since the war ended. Between 1946 and 1950, the price of oil at the wellhead had risen from $1.50 to $2.37 in Californian fields, from $1.70 to $2.65 in East Texan, from $1.62 to $2.57 at Midcontinent, and $3.55 to 4.10 in Appalachian.[94] But the tide was turning in the market. New technologies were stimulating remarkable improvements in the ability of oil companies in the upstream. Offshore production came on stream during the close of the 1940s. Wells drilled from fixed platforms in stable, shallow waters were nothing new—that had been done in Louisiana and Venezuela already. But Kerr-McGee's venture 10.5 miles off the coast of Louisiana was indeed new. When they struck oil, their incredibly risky venture paid off, enticing their competitors to follow suit and head offshore in a completely new way. Natural gas emerged as a potent new energy source in these years as well. Transporting the stuff had been a challenge in the past, but thanks to the construction of two major pipelines in 1947 leading to major consuming areas in the Northeast, the gas trade took off. By 1950, the interstate gas trade had reached a flow of 2.5 trillion cubic feet—an astounding 2.5 times what it had been in 1946. This increase in natural gas flow curbed the demand on oil significantly. One estimate judged the additional volume of natural gas as lessening oil demand by as much as 700,000 barrels per day.[95]

By 1950, the situation in the domestic market was unrecognizable to those who had fretted over supplies just two years earlier. Domestic crude oil production jumped from 4.7 million barrels a day in 1946 to 5.4 million in 1950. Proved domestic reserves of crude oil and natural gas liquids jumped by more than 2 billion barrels in 1948, the biggest one-year jump since 1937.[96] Record distillate inventories accumulated in the winter of 1949, 50 percent greater than they had been the previous year.[97] J. C. Donnel, president of Ohio Oil Company, noted the dramatic reversal of circumstances at the company's shareholders' meeting in 1949: "Last year at this time . . . the industry was straining its facilities to meet rising demand. By late fall, however, capacity operations, output and stocks were at high levels and the shortages of petroleum were history."[98]

The panic about shortages present and future ceased as a tidal wave of oil, much of it from vast new fields being exploited in the Middle East, came crashing into the United States. Yet those who had predicted a rising tide of foreign imports proved prescient. These imports had helped to ease the shortages in the domestic market, containing prices. Net crude imports began to grow dramatically: 118,000 barrels per day in 1945 grew to 391,000 barrels per day just five years later. By 1959, imports would reach just shy of one million barrels a day. The role of foreign oil in total petroleum supply would grow as well during these years, from 2.5 percent in 1945 to 6.2 percent in 1949, eventually nearly doubling to 11.9 percent by 1959.[99] If anything, the challenge facing the industry at the dawning of the new decade was one of overabundance, not scarcity. The old anxiety about demand for oil growing more quickly than production could possibly keep up with evaporated.

In 1950, *Time* looked back at the oil fears of the previous years, noting how times had changed: "The U.S. oil industry thought a year ago that its longgushing boom was trickling out. In its hell-for-leather expansion, the industry has overtaken all demand and for the first time since World War II found itself producing more oil than it could sell."[100] It was a happy—if temporary—state of abundance that would last for more than two decades.

4

"A Problem Unprecedented in Our History"
American Anxiety in the Age of OPEC,
1970–1986

The decade of the 1970s remains a time in American history almost synonymous with mayhem in the oil market, and deservedly so. The spare production capacity of American oilfields—that critical buffer between how much they were producing and the absolute ceiling on what they could produce at a moment's notice to stave off a spike in prices—would disappear by 1972, leaving the erstwhile most powerful producer in the West in a position of vulnerability it had never known. A handful of key producers, sensing their own geostrategic import to the West and tired of having to play by the rules of the pricing game as dictated by oil majors, formed the Organization of Petroleum Exporting Countries (OPEC) in 1960, eventually prompting a wave of nationalizations and production reductions. Expert predictions of a looming "energy crisis" first began to pop up in 1969 but soon grew more common, from the dire warnings of the U.S. State Department to the gloomy prophesies of the Club of Rome. Amid these tensions, two genuine supply crises in the market, brought about by political conflict, made talk of a protracted oil shortage possibly right around the corner reach a fever pitch, first in 1973 and later from 1979 to 1980. Predictions of an imminent, irreversible shortage of oil supply became ubiquitous. Surveys suggested that the American public was increasingly under the impression that the country was "running out of energy." To many Americans, the Iranian Revolution and the onset of the Iran-Iraq War in 1979 and 1980 appeared to be the fruition of the foreboding predictions made throughout the preceding decade. But this wave of fear, too, would pass—and in dramatic fashion—only half a decade later.

The landscape of oil in the United States—the players, prices, politics, and policies that had defined the industry for many years—changed radically during the 1970s. The preceding decade had been one of economic boom, thanks in no small part to a wave of cheap oil. Oil consumption outside the

Soviet Union ballooned from 30 million barrels per day in 1965 to 45 million barrels per day in 1970. It would leap to 55 million per day just three years later.[1] Oil was being used to create a growing array of products, whether in plastics or in textiles, and the world economy, from the power plant to the gas tank, was becoming more and more oil-dependent. Globe-straddling American and European oil giants, firmly in control of the world's largest oil fields, masterfully managed the supply side of the market to their own benefit. These companies, derisively termed the "Seven Sisters" by the Italian oilman Enrico Mattei, were vertically integrated powerhouses with the political backing and the financial heft to administer the global industry. But their position astride one of the world's most powerful industries, with unrivaled dominance, was not to be permanent.

"The Winds of Change … Have Risen to Hurricane Proportions"

For American consumers, as well as for the companies that pumped the oil, the late 1950s and 1960s was a golden age. Cheap oil stretched out as far as anyone could see. Reflecting on this halcyon era in the oil market, one expert described the heady self-confidence that defined forecasts about future oil prices during that decade in a 1973 *Foreign Affairs* article:

> It was the popular, almost universal theory of the 1960s—still vigorously defended by a few of its early proponents—that this abundant supply of oil, whose cost of production was very low, and which was found in all corners of the earth, would soon be sold at its "proper" economic price—apparently $1.00 per barrel or less—and for some time it was confidently predicted that this price would prevail in the Persian Gulf in 1970.

There was no better evidence of this widespread self-assurance that prices would stay low indefinitely, and that the United States would have little need for large-scale imports any time soon, than the findings of President Nixon's Task Force on Oil Imports in early 1970. The task force, headed by Labor Secretary George Shultz, saw more good times ahead. Prices would rise only modestly through 1980. Of the 18.5 million barrels per day the country would consume, it projected all but 5 million barrels per day would be produced domestically. Most other imports would come from the Western Hemisphere. This rosy picture was not to be, of course. Imports would jump to 6 million barrels a day by 1973, well before the 1980 timeline the Task Force envisioned, and crude oil prices would reach levels that Nixon's luminaries could hardly have dreamed possible. But the report was not the opinion of a few isolated academics. Its findings were informed by the work of the nation's largest oil companies, the National Petroleum Council (an advisory committee first convened by Truman in 1946 to give the White House the industry's views on the oil and gas business), and the Department of the Interior.[2]

One of the first ruptures in the world oil market that would occur in the 1970s likely went unnoticed by nearly everyone in America outside the oil patch, but it would set the stage for the oil crisis that would strike just a few years later. By March 1971, U.S. oil fields, once able to produce far more than they actually were, were producing all-out. The country had lost all of its so-called spare production capacity, or the ability to quickly ramp up its oil output within the space of a few months for an extended period. This was a massive change. For the first hundred years or so of the U.S. oil industry, the country had always been able to produce more oil than it needed. In fact, overproduction was an ever-present threat to domestic drillers in times of waning demand, who ran the risk of glutting the market. It had happened during the Great Depression as it had periodically before that. Although Rockefeller's Standard Oil was regarded a villain at the time, one of the often-overlooked benefits of his self-interested stewardship of the domestic market was that he prevented prices from collapsing. After Standard Oil was broken up, various regulatory bodies, such as the Texas Railroad Commission and the Oklahoma Corporations Commission, eventually had to step in, putting in place their own prorations on output so that prices stayed high to prevent waste and keep the domestic oil industry afloat.

A combination of factors was behind the loss of U.S. spare capacity at the outset of the 1970s, a change that would carry important implications for prices and the geopolitics of oil. On one hand, domestic oil consumption was rising briskly, up 25 percent in the prior five years alone. Meanwhile, crude prices were at rock bottom. In today's dollars, they never rose above $14 per barrel in terms of average annual prices between 1961 and 1971.[3] No other country in the world had been drilled over to the extent that the United States had. Other, more virgin fields in the Middle East and North Africa, for example, could be produced much more cheaply, which meant that they received the lion's share of new investment, and import quotas were bringing the oil into the domestic market at a brisk rate (though the quotas, established by Eisenhower in the wake of World War II and intended to limit the nation's dependence on foreign oil, were phased out once the nation's surplus capacity had disappeared).

With the loss of surplus capacity in the United States, the security cushion of ready oil production that Washington could call on if need be had vanished.

There was no more safety margin in the event of an emergency, whether it came in the form of a supply disruption or a wartime spike in demand. The United States, and with it the Western world, would rely increasingly on the oil of the Middle East to supply growing demand, a fact that would eventually play a role in the creation of the U.S. Strategic Petroleum Reserve. There would be no potential for domestic wells to make up for an unforeseen jump in consumption. The country's oil economy, now reliant on abundant supplies of cheap energy, was in a position of vulnerability like it had never been, and one that other oil producers quickly learned they could use to their advantage. It was now the Middle East, once an afterthought to American politicians, that would assume the top of oil's pecking order.

Relations between Western oil majors and Middle Eastern powerbrokers had been worsening for some time by 1970. The new oil titans of the Middle East and North Africa, increasingly aware of their central place in the world economy thanks to their abundant easy-to-access oil reserves, chaffed at the oil companies' practice of "posting" oil prices—more or less announcing their offering prices to buyers based on their own read of market fundamentals, with little input from the sovereign government whose land they drilled. The companies' decisions to raise or lower prices in a given country could be of huge importance to its economy. Low posted prices could mean a dire drop in public revenues. The industry's nasty habit of producing too much oil, which invariably pushed prices downward, had depressed crude prices from the mid-1950s through the 1960s, cutting deep into the producing nations' revenues.

A new generation of nationalistic leaders in the leading exporting nations in the 1950s and early 1960s, such as Libya's Muammar Qaddafi, among others, tapped into public discontent over the perceived abuses of the oil multinationals. Rather than play by the established rules of the game—courting oil company investment and supporting the concession system—they had a different idea for how things should be run. The oil was theirs, not the West's, they argued. And the oil companies and Western governments were all too synonymous, as they saw it. The oil belonged to the people of their nations. Its ownership and exploitation was a matter of national sovereignty, not mere business. And they would do what was needed to restore control of the industry to its rightful owner—those governments blessed with oil in the ground.

Frustrated by a series of downward revisions to the posted price during the closing years of the 1950s, the governments of some leading oil-exporting countries finally decided to act. When Exxon reduced its posted price once again in September 1959, bringing the price of Arabian Light to $1.91 per barrel, the other multinationals followed suit. They would drop prices by an additional 10 percent in August 1960.[4] Oil exporters had had enough. Juan Pablo Perez Alfonso, the Minister of Mines and Hydrocarbons in Venezuela, concluded that it was high time that the Western oil companies that held such sway over his country's major export product were brought low. Perez Alfonso had authored Venezuela's influential fifty-fifty profit-sharing formula some years before, which had increased the country's take of oil revenues to some extent. But the time had come, he concluded, for more drastic measures if the global price of oil was to be stabilized.

In September 1960, he assembled representatives from Venezuela, Saudi Arabia, Iraq, Iran, and Kuwait in Baghdad to discuss how their governments could band together to take greater control over the oil trade, and in so doing, over the long-term future of their national economies and public finances. In the course of the meeting, OPEC—the Organization of Petroleum Exporting Countries—was born. According to its founding vision, OPEC would serve as an instrument for collective bargaining and defense. It would reclaim from the oil companies the primary role in the decision-making process and management of oil activities in their territories. It would aid the group's effort to work together with the oil companies as a united force to negotiate prices, production, taxation, and policy.[5] Finally, it seemed, the chief exporting nations had taken control of the means to stabilize prices and production over the long term, which would allow them greater control over their long-range fiscal position and economic development strategy.

OPEC did not flex its muscles over global oil supplies until well after its founding in 1960. In its first years, the nascent cartel was "all thunder and no storm," in the words of one historian.[6] The five countries that comprised OPEC at the outset had a harder time working together than they had envisioned in Baghdad in 1960, despite the potential for tantalizing mutual benefits if they could pull it off. Iran and Saudi Arabia, bitter rivals in the oil trade, vied for supremacy as the West's most powerful ally in the Persian Gulf. Kuwait was eager to get a greater share of the pie. Internal jealousies led to squabbles that undermined the effectiveness of the infant institution. But the OPEC's founding five did not lose sight of their initial vision, waxing increasingly bold in their dealings with the majors in the 1960s and early 1970s.

Momentum seemed to be on their side. A single OPEC country's success at renegotiating the terms of a production contract with a foreign company inevitably set off a domino-like effect on its peers, who would in turn demand an equal or better deal. That meant that a nationalization or contract revision forced upon a company by one government usually led to better terms for its fellow oil exporters. By mid-1971, OPEC members had signed agreements with the oil majors that ensured them all 55 percent share in oil profits and a higher posted price for all producer countries. In a few of the countries, nationalizations were not far away. Algeria had expropriated 51 percent of all foreign company's hydrocarbon holdings by 1971. Qaddafi seized 51 percent of BP's assets in Libya that same year. Then Iraq, following Libya's lead, expropriated the Iraq Petroleum Company's assets in 1972.[7] The game had changed for the majors.

Walter Levy, the former head of the petroleum section of the Office of Strategic Services, the forerunner to the CIA, described the situation in *Foreign Affairs* in the summer of 1971: "The economic terms of the world trade in oil have been radically altered. The balance among oil-producing and oil-exporting countries and oil-consuming and oil-importing countries, and among oil companies themselves appears, at least as of now, to have shifted decisively in favor of the producing countries." Put more dramatically, "The winds of change" that had been "stirring throughout the decades since 1950 have now risen to hurricane proportions."[8] The Western oil majors—squarely in control of the world oil trade going back several generations—were now on the run.

This rolling back of the Western energy architecture would have been dramatic under any circumstances. But given the supply and demand picture of the early 1970s, its unsettling effect on global markets was all the more profound. World energy demand was climbing, up 4 percent a year between 1970 and 1973. Coal, natural gas, nuclear power, and hydroelectric power had difficulty growing quickly enough. Oil appeared to be the most promising source of energy to fill the supply gap, but production would need to grow worldwide at a record pace of roughly 8 percent a year to do it. Experts struggled to see where such astounding increases in output or new sources of supply capable of filling the gap could be found. Oil production in the United States was clearly not up the task: crude output in the lower forty-eight states had peaked in November 1970 and was trending lower and lower by the year. The Soviet bloc, once one of the world's great oil-exporting regions, appeared poised to become a net importer, a flip that troubled Western energy officials. The Soviet Union's unwillingness to disclose data on its energy production and usage only made the picture all the hazier and more unpredictable.

The oil industry was moving at a breakneck pace to try to accommodate the growing thirst for energy. Investment in every link along the supply chain, from the wellhead to the gas station, was happening on a tremendous scale. Between 1970 and 1973, it rose by almost 50 percent, more than double the rate of the previous ten years, which had been a period of substantial growth in its own right. Upstream investment was highly concentrated in the Middle East, the epicenter of supply growth in the 1960s and early 1970s, up 52 percent between 1970 and 1973 alone. New refining capacity was also growing briskly. Over seventy new refineries were being constructed and 120 were being expanded around the world in 1970. By the end of 1975, world refining capacity was 30 percent higher than it had been five years earlier. The world's fleet of oil tankers grew by 22 percent higher in 1970 and 1971 alone.[9] Oil was fueling the engine of economic growth in the early 1970s. And it was the Middle East—now synonymous with OPEC, for all practical purposes—to which the world was looking to meet ever-rising demand.

"A Permanent Sellers' Market"

Many analysts in the oil industry were skeptical that enough oil could ever be found to meet the world's demand within a decade and a half. In July 1969, the *Petroleum Press Service*, the forerunner to the *Petroleum Economist*, noting that the "demand for oil" was still "vigorously expanding" as the 1960s drew to a close, and could possibly even double between 1968 and 1980, worried aloud about the Herculean task of finding enough oil to satiate such a thirst. "The petroleum industry is faced with the task of finding large new reserves, and it must make continuous large-scale investments in new production, refining, transportation and distribution" or risk falling behind demand growth.[10] In April 1971, BP's chief geologist, H. P. Warman, expressed skepticism that the industry could perform the "daunting task" of finding the 900 billion barrels of oil he believed it would need before 1985 to replace quickly depleting reserves. The U.S. National Petroleum Council estimated that oil consumption in the free world would rise from 37 million barrels per day in 1970 to 92 million barrels per day by 1985, a seemingly impossible demand to meet. The United States would need to import as much as 300 percent more oil as a result over that time frame. Walter Levy's consulting firm warned that world demand was set to increase by 35 million barrels per day by 1980—an 80 percent jump from 1970. Shell and Texaco both warned that world oil demand would nearly double, from 40 million barrels per day in 1970 to between 70 and 79 million barrels per day in 1980, and that the United States would need to import six times more oil in 1985 to meet domestic demand.[11]

These forecasts were typical of a generation of oil market projections in the 1970s, almost all of which contributed to unease among the oil industry, national governments in both oil-exporting and oil-importing countries, and the general public over whether oil would be a viable energy source in several years' time. Opinions varied as to its seriousness, though only some observers seemed to think it was a serious problem.[12] Demand for all forms of energy was growing briskly—that was an accepted fact. But few could see where the supply would come from. In the United States, federal energy regulations were making things worse. Price controls on oil at the wellhead and burner tip, which Nixon had imposed in 1971 to curb inflation, had the unintended consequence of discouraging oil exploration and production, encouraging consumption of unsustainably cheap oil and catalyzing imports. Even the elements seemed to be conspiring to make the energy situation worse. The winter of 1969–1970, the coldest in thirty years, had resulted in utilities being forced to interrupt service due to oil and natural gas shortages. Additional new demand for low-sulfur oil, which was needed to fuel a growing number of electric utilities that had switched from coal- to oil-burning generators, meant that imports from Libya and Nigeria, producers of low-sulfur oil, had to increase. The following summer, in 1970, brownouts were common along the Atlantic Coast. The regional electricity infrastructure was simply incapable of providing enough power to meet demand. Talk of an energy shortage was no longer an academic matter; it was now a public issue, gaining the attention of the average American consumer and moving to center stage as a national news item.

Among the national figures most vocal about a "looming storm" about to wreak havoc on the American energy landscape was James Akins, the State Department's Middle East expert and later advisor to Presidents Nixon and Carter. Akins had been one of Washington's leading lights on oil issues for several years, having assumed the department's top energy post in 1968. He would be appointed the U.S. ambassador to Saudi Arabia in 1973. A prolific commentator on international affairs, both in his role as a foreign service officer and as a public intellectual, Akins's forecasts about the looming calamity awaiting U.S. oil markets provide a valuable "window" on the "ideas and assumptions governing U.S. polities toward the oil markets" at the time, according to Morris Adelman, an economics professor at MIT and one of the era's most provocative thinkers on oil. According to Adelman, Akins's writings on the oil market during the 1970s offer a glimpse into what State Department elites "thought and assumed" about where oil production and prices might be headed.[13]

When it came to the future of oil, Akins was hardly an optimist. Writing in 1971 as the head of the State Department's Office of Fuels and Energy, he argued that the country was entering a never-ending era of shortage. Washington was enjoying the "last gasp in the buyer's market" for crude oil. In terms of supplying itself, the best the United States could hope for was to extract more oil from the Outer Continental Shelf and squeeze whatever liquids it could from kerogen and coal, which would be possible only at a high cost. "By 1975," Akins wrote, "and possibly earlier, we will have entered a permanent sellers' market, with any one of several major suppliers being able to create a supply crisis by cutting off oil supplies."[14] The State Department forecast that oil prices would more than double between 1970 and 1980, rising to "a level equal to the cost of alternate sources of energy," an unthinkable reality. Akins predicted that the price of a barrel of oil would hit $5 in the U.S. Gulf of Mexico by the end of the decade. State Department officials were not shy in broadcasting their forecasts in testifying before Congress and taking their message on tour across the country. As it turned out, Akins and his team were all-too-right that oil would be more expensive in 1980 than in 1970—though his forecast, ridiculed at the time as "alarmist and provocative," would prove far too modest. Benchmark Arabian Light crude oil in the Persian Gulf would average $36 per barrel in 1980—almost an order of magnitude higher than anything Akins ever contemplated.[15] Notwithstanding the correctness of his call on oil prices, however, it would not be accompanied by the "permanent sellers' market" he envisioned.

Akins and his colleagues continued to sound the alarm in 1972, as world benchmark oil prices busted through the $2-per-barrel barrier, then $3 per barrel. The country would soon find itself in a position that was "nothing short of desperate." Washington needed to wake up to the fact that oil is "the most political of all commodities. We must recognize it because the Organization of Petroleum Exporting Countries recognizes it," citing a Libyan official who had recently declared that "we must brandish . . . the oil weapon . . . to enable us to win this battle." The notion that the United States would have access to "cheap foreign oil" in the future was erroneous, even dangerous.[16] John Irwin, Undersecretary of State, warned the leadership of the Organization for Economic Cooperation and Development in 1972 that "they all faced a severe shortage of energy supplies by 1980." All told, the industrialized world may require as many as 26 million additional barrels per day by that time, an increase roughly the size of U.S. and Western European consumption at the time. The question of where the oil would come from, if OPEC was not willing to share, was frightening.[17]

This Time the Wolf Is Here

But it was Akins's alarming piece that appeared in a 1973 issue of *Foreign Affairs* that got the most attention among American elites. Its title left little room for doubt about where Akins stood: "The Oil Crisis: This Time the Wolf Is Here." He acknowledged that oil's prophets of doom had indeed been wrong in the past—but this time really was different:

> Oil experts, economists and government officials who have attempted in recent years to predict future demand and prices for oil have had only marginally better success than those who foretell the advent of earthquakes or the second coming of the Messiah. The recent records of those who have told us we were running out of petroleum and gas are an example. Oil shortages were predicted again in the 1920s, again in the late thirties, and after the Second World War. None occurred,

and supply forecasters went to the other extreme: past predictions of shortages had been wrong, they reasoned, therefore all such future predictions must be wrong and we could count on an ample supply of oil for as long as we would need it.

Such optimistic forecasts were naively wrongheaded, Akins argued. The United States and its allies were becoming dangerously dependent on Middle Eastern oil, assuming that it would continue to grow strongly. If the oil failed to materialize, be it for geological or political reasons, "there could be a real supply crisis in the world." It did not bode well, Akins presciently observed, that in 1972 alone "Arabs in responsible or influential positions made no less than fifteen different threats to use oil as a weapon against their 'enemies,'" and Washington was enemy number one. What did all this mean for the future of oil? Could it last as a viable source of energy for the United States—or for that matter, for mankind? It was impossible to "look simply at the world's oil reserves and conclude that they [were] sufficient to meet the world's needs" in the years ahead. Such "fatuities" had to be permanently disregarded as childishness, Akins wrote.

Yet there was reason—albeit dark—for hope. The energy crisis the world would soon face "will not be a long one in human terms." As reserves waned, prices would rise, setting off an international crisis with no obvious end game, forcing consumers to get their energy from other sources. "By the end of the century," Akins predicted, "oil will probably lose its predominance as a fuel." Solar and geothermal energy would be part of the solution. Others sources, like nuclear fusion, which he admitted as "purely hypothetical" for the time being, might also need to come to the rescue. But even these sources of energy could not be developed quickly enough at scale to save the United States from the crisis that was to come, given how long it would take to bring alternative energy sources online. No white knight would rescue the industrialized world from the gathering storm.

Akins was not alone in his pessimism about the trajectory of the world's natural resource economy. It was in the mid-1960s and early 1970s that the modern environmental movement, a groundswell of popular support in the industrialized world for rethinking the way humans related to the natural world, was born. Environmental issues were making their way to the forefront of American political consciousness, challenging deeply held views about the sustainability of modern industry and drawing attention to forces that advocates saw an anti-environment. At the top of the list of these no-good-doers was Big Oil, a fact that has changed little in the intervening forty years. Grassroots efforts to make public policy more environmentally friendly were becoming more common, popular, and politically powerful. Rachel Carson's bestselling book, *Silent Spring*, published in 1962, helped sow the first seeds of the movement. Linking chemical companies to environmental degradation and deteriorating public health, Carson lambasted public officials for their easy treatment of big business in pursuit of the almighty dollar at the expense of the rest of the world's quality of life.[18] In 1967, the U.S. Senate overwhelmingly passed a clean air bill, following New York City's lead in limiting coal consumption in urban settings. Three years later, federal legislation mandated the use of environmental impact statements before new ground could be broken on any major development projects. Earth Day, founded by Gaylord Nelson, a U.S. senator from Wisconsin, was initiated in 1970. Borrowing the tactics of the antiwar effort, Nelson helped launch a bipartisan campaign to try to "force environmental protection onto the national political agenda," bringing 20 million Americans to the country's streets, parks, and auditoriums on April 22 to demonstrate their commitment to a cleaner future.[19]

The Coming "Global Minerals Shortage"

No publication better encapsulated (and fed) the rising alarm among elite thinkers and the American public about humankind's perilous prospects than *Limits to Growth: A Report for the Club of Rome's Project on the Predicament of Mankind*, published in 1972.[20] Drawing on what was at the time a stunningly advanced computer simulation developed by researchers at MIT, *Limits to Growth* purported to peek into the future of the global ecosystem using an impressive array of mathematical equations that extrapolated out historical trends. The Club of Rome's simulation wove together what it considered probable outcomes along five interrelated variables—industrialization, pollution, food production, mortality and fertility rates, and resource depletion (including oil and natural gas)—from 1900 until 1972. Its goal was to simulate how these variables might interact over the next century and beyond, and finally rendered a forecast of the state of the world in 2100.

The picture it painted of the future was sobering. According to the simulation, the rapid depletion underway of vital resources like oil and coal would leave nonrenewable energy sources all but extinct by 2050, by which time exponential growth in pollution and rapidly declining food availability per capita would result in an explosion in the death rate. Oil would be hard to find by the mid-1990s. Then, after the human population had skyrocketed to an unsustainable peak by 2030, the number of people on the planet would plummet as the world plunged headlong into its "limit to growth." Industrial civilization and the abundance of modern life would fade away to very low levels sometime between 2050 and 2100, perhaps never to return. A new stasis for the human population would settle in, defined by an almost-complete void of nonrenewable energy resources and little food production. Humankind's only hope was to curb its population and industrial growth severely—otherwise, the world faced the prospect of a catastrophic crash. Finding more oil or other resources would only delay the inevitable day of reckoning.

The study did not predict the exact timing of future crises, but it did warn of the dangers of a hydrocarbon-based economy as unsustainable and soon to be exhausted. In hindsight, the bedrock assumptions about the future of oil supply and demand, however, were deeply flawed. *Limits to Growth* all but ignored the possibility that high prices for critical resources would push consumers toward substitute goods. It lazily lumped all exhaustible resources, from conventional crude oil to sedimentary phosphate rock, into one giant category, without allowing for the dynamic interplay of supply, demand, and price among this extraordinarily diverse group of commodities. The report also unrealistically assumed that the world's remaining oil reserves amounted to 550 billion barrels and the compound growth rate of consumption would grow at a relentless 4 percent a year until 1992. Neither assumption was plausible. As it turned out, consumption increased at less than half the projected rate, even as the world produced more than 700 billion barrels between 1972 and 2004. Even after those billions of barrels had been pumped, proven reserves would still exceed 1.38 trillion barrels in 2010.[21] Oil prices would rise sharply at intervals during the 1970s, and again after the turn of the century, but never in the relentless upward march into the future that the Club of Rome envisioned. Commenting on its inaccurate forecasts decades later, one scholar described the once-sensational *Limits to Growth* as an "exercise in misinformation and obfuscation rather than a model delivering valuable insights."[22]

What was most unnerving to everyday people about *Limits to Growth* was that its dire predictions appeared to be coming true. Just as the Club of Rome had predicted, energy sources appeared to be getting scarcer. American oil production outside Alaska could not be sustained after its 1970 peak. The country's oil imports were shooting through the roof, hitting new records year after year, even as demand for oil globally was climbing. Meanwhile the world's population growth showed no signs of slowing down. This was all just as the authors had foreseen. *Limits to Growth* seemed increasingly prophetic in the years that followed its publication, apparently borne out by what appeared to be unstoppable economic, environmental, and demographic trends. Its vision of the future became part of the zeitgeist of the 1970s, injecting a "potent element in the fear and pessimism about impending shortages and resource constraints that became so pervasive in the 1970s, shaping the policies and responses of both oil-importing and oil-exporting countries," as one historian would later describe it.[23]

What was more, other official forecasts at the time corroborated its conclusions. The director of the U.S. Bureau of Mines, speaking in September 1973, predicted that by the end of the century, the United States "will be competing for those foreign mineral products," including crude oil and natural gas, "with other mineral-hungry nations all over the world. The situation could lead to a global minerals shortage that would make our current energy crisis look like the good old days by comparison."[24] As he saw it, the wolf was indeed at the door. It was just a matter of when he would decide to let himself in.

"Market-Oriented Prices Vanished" Amid "Extraordinary Fragmentation and Volatility"

Even before the fireworks in the oil market began going off in the final months of 1973, the oil market had been unusually jittery. Between December 1970 and the summer of 1973, official oil prices for Arabian Light crude more than doubled. Even the previously simple job of tracking and publishing price movements in the oil market—the work of various petroleum journalists and industry newsletters—began to prove vexing to industry analysts, so schizophrenic was the market compared to what it had been since World War II. By August 1973, global market prices were all but "imprintable," as the *Petroleum Intelligence Weekly*, or *PIW*, the industry's go-to guide for intelligence in a shadowy industry, described it. "Accurate and conscientious oil price reporting has never been easy," the newsletter observed. "But it's getting next to impossible in the current 'disintegrating' market . . . what with flocks of inexperienced newcomers coming like lambs to the slaughter." Differentials among the prices for various refined products, spot barrels, and other discounts and pricing schemes began to oscillate wildly. The "upward pressure" on prices was extraordinarily intense, as the *PIW* described:

> Put all these rubbery numbers and muddied psychological factors into your crude price calculations, and what you come up with is nothing concrete price-wise, but the sure knowledge that there's firm upward pressure on the price of the crude that is being offered by state sellers everywhere.[25]

The days in which the Seven Sisters had seamlessly managed oil prices within a carefully defined range seemed long ago. Now, in sharp contrast, the market had become one of "extraordinary fragmentation and volatility."[26]

For the first time, the price that a barrel of oil was commanding for immediate delivery, or spot prices, diverged sharply from the oil companies' official posted prices for weeks and months at a time. Oil refiners' fears of the possibility of a shortage of oil on its way caused them to bid up the price of any extra crude on the market. As a result, more and more oil—though still only a small percentage of total transactions (less than 3 percent)—was being purchased on the spot market rather than through long-term contracts at official posted prices. The price of spot crude jumped above $5 per barrel, even $6 per barrel, at a time when posted prices were never higher than $3.[27] In August 1973, *Petroleum Intelligence Weekly* reported that "near-panic buying by US and European independents as well as the Japanese" was pushing spot prices higher and higher. The exploding spot market was making the oil producers all the more confident about their position. OPEC, well aware that spot crude prices were commanding a premium relative to the posted price, took spot prices as a signal that perhaps their posted price was not high enough after all.[28] If consumers were ready to pay well above official prices, couldn't official prices go higher without spoiling demand—the argument that the Seven Sisters typically made for why OPEC should keep prices low? Perhaps, the leaders of oil-rich Middle Eastern countries concluded, they could afford to be bit bolder about dictating prices than they had been.

Meanwhile, trouble was brewing in the Arab world as the midway point of 1973 passed. At the Munich Olympic Games in August 1972, members of the Palestine Liberation Organization (PLO) took several members of the Israeli Olympic team hostage in a shocking and highly publicized act of terror. The old wounds of the 1967 Six-Day War, still festering, had spread violent anti-Western radicalism across the Middle East. A new Egyptian leader, Anwar el-Sadat, was actively conferring with his peers in the region about launching a new military campaign against Israel to recapture the territory it had taken from Cairo in 1967. Sadat was convinced that the chances of the Arab states, banding together, being victorious over Israel in an armed conflict would be much higher if they were to wield what he felt was their strongest source of leverage against the West: the "oil weapon," as he called it. He knew that the United States, a staunch ally of Israel, would likely come to its aid in a war against one or more of the Arab states. The threat of cutting off oil supplies to the United States, he was convinced, could keep Nixon and Kissinger at bay. All he would need to wield the oil weapon was the agreement of the key Middle Eastern oil producers to boycott oil imports to the United States should it interfere in the war against Israel. In May 1973, King Faisal of Saudi Arabia ensured Sadat that his country would cut off oil supplies to any Western country helping Israel. Syria had already agreed to provide Egypt with military aid.

On October 6, 1973, war broke out between Israel and an alliance of Egypt and Syria. It would become known as the Yom Kippur War. Golda Meir, the Israeli prime minister, pleaded with Nixon for military assistance for her country on what appeared to be the verge of collapse. He responded, sending what he intended to be a secret airlift of munitions to Israel. The plan failed, however, when the planes were spotted. OPEC's threat to boycott the exportation of oil to any country that aided Israel in the war had been triggered, and it responded quickly.

On October 16, a delegation of Persian Gulf countries—Saudi Arabia, Iran, Iraq, Kuwait, the United Arab Emirates, and Qatar—all OPEC members, announced a unilateral increase in the benchmark price of Arabian Light crude from $2.90 to $5.11 per barrel.[29] The whopping 76 percent price hike shook the oil industry the world over, but for more than just the higher price. OPEC's October announcement marked the unofficial end of the posted price system, whereby the West's most powerful oil producers could set the price of a barrel of oil. The practice of setting prices, begun by Rockefeller many decades earlier, was over. But OPEC was just getting started. The following day, members of the Organization of Arab Petroleum Exporting Countries (OAPEC), a regional subset of OPEC, announced that it would begin lowering its production volume by 5 percent every month until Israel relinquished to Egypt the territory it had occupied since the Six-Day War in 1967, effective immediately. Only states that supported Israel would be subject to those cuts in production. It was a "selective embargo" against Israel and its allies, and Nixon's airlift of war supplies to Israel the week prior meant that the United States would ineluctably be subject to it.

The effects of the announcement on the oil market were convulsive and immediate. Prices flew through the roof. By the end of 1973, Libyan crude was reported to be commanding $9 per barrel, whereas it had only garnered $2.80 a year earlier. Iran lifted the price of its November crude to Asia from $2.88 to $3.85 per barrel. Iraq raised the posted price of its crude to around $9 per barrel. A November 16 auction of spot Tunisian crude sold for $12.64 per barrel. No one knew what the exact market price of crude was, but no company that depended on it to do business—whether an airline, an energy brokerage, or a refining operation—and did not have sufficient emergency inventories was going to go without it, so they kept buying.

Still prices pushed higher. Tehran sold 450,000 barrels of crude for $17 per barrel at an auction in mid-December. Nigerian and Algerian crude passed the $16-per-barrel mark that same month, while a new round of Iranian sales surpassed $17 per barrel once again. A Japanese trader, under heavy pressure to supply its home market and in competition with over eighty other buyers, bid $22.60 per barrel at a Nigerian auction. The deal was never consummated, but news of the failed deal made the rounds. OPEC, unfazed by the skyrocketing prices despite their potentially dampening effect on longer term demand, decided in December to double the posted price of their benchmark crude. It was now $11.65 per barrel. In light of posted prices only a short time before, it was a staggering development. The price was four times higher than the posted price only four months earlier—and a full ten times higher than the 1970 price. The new price would go into effect January 1, 1974.[30]

Any certainty over the future of oil—its price, the quantity that would (or even could) be supplied down the road, the true extent of the embargo—was gobbled up in the frenzy. In one observer's words:

> Talk of market-oriented prices vanished. No one knew what "the market" was, or where it was going, or how it could serve as a guide for the cartel's price. In fact, the market was still thin. It was not institutionalized. There was simply no solid reference point, just a bunch of disparate deals erratically reported in the trade press. People's knowledge of the market was confined to the deals they did themselves, what they read in the trade press (not necessarily accurate and almost always with insufficient detail), and what they could glean from their business and personal contacts (information to be treated with great caution).[31]

Rumors and stories—some true, others not—about the price points of oil products for various transactions flew among traders, executives, producers, consultants, brokers, and the media.

Some of the demand for oil was precautionary, bought by refiners and others who wanted to store it as a sort of self-insurance policy against higher prices or outright shortage later on. No one knew how bad the crisis would get, and no one wanted to be caught on the wrong side of the shortage. The crisis was "driven by the fear of dearth" by those consumers who would suffer if future production did not meet their needs.[32] There was a self-fulfilling element to the market's worries over a rapidly worsening shortage. Amid the uncertainty over whether tomorrow's supplies would be enough to get them by, buyers sought to get their hands on more oil than they would normally need, storing it against the threat of a worsening shortage.[33] As buyers entered the market en masse, prices jumped. Rising prices and unusually large demand only increased the perception in the market that something was going terribly wrong.

Other demand for oil was more speculative. The fact was, OPEC stood to gain from taking prices higher by cutting back on volumes, given that most of their buyers would not be dissuaded even at previously unthinkable prices. OPEC knew that it was earning more money by selling less oil. Taking that into account, some market participants began to fear that OPEC might keep production down permanently—it appeared to have an incentive to, at least until people found ways to use less oil. Those who bought oil had reason to worry that the embargo was merely the first chapter in an era of permanent shortage and, hence, higher prices. U.S. Secretary of State Henry Kissinger, in December of 1973, described the "crisis" as much more than "simply a product of the Arab-Israeli war." Rather, as he understood it, the shortage was the "the inevitable consequence of the explosive growth of worldwide demand outrunning the incentives for supply."[34] Seen through this lens, not all "panic buying" was as irrational as the word "panic" would suggest. Rather, from the standpoint of a speculator, it was a directional bet in favor of the possibility of higher prices later on, at which time the contracts or even the physical oil could be resold.[35] Were OPEC to raise the price again, speculative oil bought at the close of 1973 might have been sold at a sensational profit. Fear of a new era of greater scarcity was moving the market in a gravity-defying way.

The denouement of the embargo turned out to be more mundane than its dramatic beginnings. The plot went out with a whimper. On December 4, the Saudis announced that they would no longer go forward with the month's scheduled 5 percent production reduction. They gave no reason for their refusal to continue their participation in the embargo, but some speculated it came as retribution for Iran and Iraq's failure to cut back their production.[36] By the latter half of December, the tension was easing. The possibility of shortage, the subject of intense worry only weeks earlier, appeared to be dimming. The *New York Times* reported on December 29 that oil tankers were lined up off the coast of the major refining ports of the Atlantic seaboard, awaiting their turn to unload. By early 1974, the production cutbacks were over. There was never any formal announcement or official reconciliation between Israel's backers and the OAPEC ending the embargo, but those who had initiated the embargo made it known to industry insiders that they were abandoning the cutbacks. Arab oil ministers, meeting in Kuwait on December 25, now called for a 10 percent production increase in January. The conditions that the Arab countries had imposed were never met. Israel won the Yom Kippur War, largely on the back of the military support of the United States. By the first few months of 1974, it was clear that the oil crisis was over.

When the dust settled, the reality of the 1973 oil crisis was that a physical shortage of oil had never in fact occurred—"crisis" notwithstanding. There was never any true shortage between 1971 and 1974, and net losses in U.S. companies' oil in storage had been minimal at most. Oil market data was so spotty during the embargo that no one in the market could tell where supply levels actually stood, although the less-scary truth of the situation was becoming clear to some experts by late December. Exports from Arab shipping terminals were nearly 40 percent higher in the first week of December 1973 than they were a year earlier, with the data suggesting that Middle Eastern producers "may not have cut production by anything like the amount" they claimed.[37]

The full story became discernible only after the fact. There are no authoritative figures, but most estimates are in relatively close agreement about what took place. Total production in the Persian Gulf was 19.4 million barrels per day in September 1973. At the peak of the embargo, that number fell to 15.4 million barrels. The output from non-Arab producers, many of which shrewdly upped their prices to take advantage of the stratospheric market, added around 900,000 barrels per day. Thus, around 3.1 million barrels per day were taken from the world market on average during the embargo. The lost oil amounted to approximately 5 percent of daily world consumption and 10 percent of global trade volume.[38] Yet inventory figures at the end of the embargo made it clear that the effect of this lack in production failed to dent commercial inventories. Private inventories of crude oil held in the United States were all but unchanged between July and December 1973. In fact, the normal seasonal dip in storage that had been typical of the years prior was far more muted than usual. Heating oil stockpiles in the United States were much higher in January 1974 than they had been a year earlier, and gasoline stocks were all but unchanged.[39]

There was no question that the embargo had put pressure on the market. Production cuts had occurred, resulting in a net loss of oil. Gasoline stations in the United States had experienced fleeting shortages of oil to sell, which spawned the now–infamous lines of people and automobiles waiting their turn to fill up. Some specific grades of fuel had been in short supply. But the overwhelming consensus was that the primary reason for the gas lines was bad domestic economic policy, not a real lack of crude oil in any meaningful sense.[40] Price controls and allocations that had put in place by the Nixon administration years earlier complicated oil imports to the United States, jamming supply lines and making it harder for refiners and wholesalers to adjust to the disruption. Poor information and the temporary shattering of established trade networks, "overlaid by rabid and violent emotions," were at the heart of the crisis. Compounding it all, in one analysis, was that precautionary and speculative buying tended to exaggerate the impact of the loss on prices. It is impossible to tell precisely whether the endless parade of forecasts over the previous few years predicting unmitigated supply shortfalls over the next several decades fueled the panic in the market, but they certainly would have done little to ease panic buying and calm market fears.

Had the embargo been a success? Had the Arab producers effectually wielded the "oil weapon," as they had threatened to do? The embargo had indeed succeeded in raising the market price of oil—there was no doubt of that. Buyers' fears of a lack of oil set off a wave of precautionary buying, which sent the price of oil up. Then OPEC, having ignited the fear with its threats of future cutbacks, ratcheted up the posted price of its benchmark crude. Raising posted prices locked in the high prices that it had instigated through its threats. In so doing, they had placed a severe burden on the net-importing countries. But the embargo had failed to bring about the political objectives for which it had been initiated—namely, to quash foreign support for Israel and reclaim Egypt's lost territory in the Sinai.

Ultimately, what many failed to realize at the time was that a "selective embargo," as it had been conceived, was impossible to implement completely. Try as they might, the Arab allies lacked the power to cause a real shortage of oil in a group of countries of their choosing, though they were unquestionably able to cause a massive spike in prices. The reason was fairly simple. The oil trade is a complex network of producers, transporters, refiners, and retailers, woven together by a sleepless horde of brokers and traders. Yes, producing countries were able to take oil off the world market, causing inventories to decline and prices to rise. But an embargo by a select, even if influential, group of producers was not able to truly stop the flow of oil to the buyers they intend to punish. U.S. refiners and other buyers drew their supplies from wherever in the world they could find them; they were neither obligated, nor restricted, to source oil from any particular producer.

In the heat of the moment, however, this fact was lost on most all American politicians. As Kissinger would reflect years later, "The structure of the oil market was so little understood that the embargo became the principal focus of concern. Lifting it turned almost into an obsession for the next five months, partly because Nixon thought that it lent itself to a spectacle that would overcome Watergate. In fact, the Arab embargo was a symbolic gesture of limited practical impact." It was true, Kissinger wrote, that "Saudi and Arab oil was not shipped to the United States," but buyers simply "shifted other allocations accordingly." The "true impact" of the embargo, he concluded, "was psychological."[41] Sheikh Ahmed Zaki Yamani, Saudi Arabia's Minister of Petroleum and Mineral Resources, would later make a similar concession. The embargo "did not really imply that we could reduce the imports of oil to the United States," he admitted. "The world is really just one market. So the embargo was more symbolic than anything else."[42] But it was a symbol that Americans would not soon forget.

"Cheap Oil Is Finished"

Among the lasting effects of the 1973 oil crisis was that it gave birth to the notion of "energy independence," a political rallying cry perennially popular among U.S. presidents that persists to this day, and to the creation of the International Energy Agency. In a November 7, 1973, radio and television broadcast, Nixon called for a new era in American energy policy based on the fact that energy shortages would shortly befall the country. He warned Americans that they must "face up to a very stark fact: We are heading toward the most acute shortages of energy since World War II." Oil supplies in the coming winter, he said, were projected to fall short by at least 10 percent—and that was before the Arab embargo had started. Revealingly, the president admitted that U.S. officials at the highest levels had already been gravely concerned about looming oil shortages even before the embargo:

> Now, even before war broke out in the Middle East, these prospective shortages were the subject of intensive discussions among members of my Administration, leaders of the Congress, Governors, mayors, and other groups. From these discussions has emerged a broad agreement that we, as a nation, must now set upon a new course.

For Nixon, the answer was clear: the country needed to escape any reliance on imported energy. "Let us set as our national goal," he declared, "in the spirit of Apollo, with the determination of the Manhattan Project, that by the end of this decade we will have developed the potential to meet our own energy needs without depending on any foreign energy sources."[43] The creation of the International Energy Agency (IEA) in 1974 was born of this effort. Formally, the IEA was an association of the most powerful net-importing countries, but the strategic rationale behind the group was an attempt at providing a counterweight to OPEC. Although only a technical forum in its early years, it would grow to constitute a formal institutional means whereby participating states managed an emergency stockpile of oil to guard against future supply disruptions.

Perhaps most important, the 1973 crisis catalyzed the emergence of OPEC as a powerhouse in most every aspect of global oil affairs. Having seen their power over the importing nations and the multinational oil companies, the OPEC nations were newly emboldened in their dealings with the oil companies operating within their borders. Kuwait nationalized the Kuwait Oil Company, a BP-Gulf joint venture that owned the Kuwaiti concession, in 1974. Venezuela nationalized its oil industry, which Exxon and Shell had dominated since the 1930s. As of January 1, 1976, it became the Petróleos de Venezuela, or PDVSA. The next domino to fall was the crown jewel of the oil world— Aramco in Saudi Arabia. The Saudi government took control of 60 percent of the company in June 1974, cautiously choosing to finalize full nationalization on a rolling basis until completion in 1981.

But beyond the dramatic changes to the world oil landscape that the events of 1973 ushered in, the crisis had caused a second alteration to the energy issue in the United States. It was perhaps more subtle but ran just as deep. It was a change in national attitude toward energy, and indeed, to American self-reliance, one of the pillars of the national ethos. The gas lines of 1973 became a symbol to Americans of a newfound sense of vulnerability, of the realization that their own livelihoods could be affected by the dictates of a part of the world that most in the United States had never paid much mind to. Unlike almost any other good, gasoline prices screamed at Americans from every street corner, incessantly bearing bad news when demand ran ahead of supply. Fluctuations in gas prices were like a little stock ticker for the national psyche. Skyrocketing gasoline prices caused public worry—not just about pocketbooks but also about the future of the nation locked in the midst of the Cold War. The findings of a public opinion survey, commissioned by the Nixon administration, revealed the depth of the strain the crisis had put on the nation's sense of well-being. "People are growing fearful that the country has run out of energy," the report found. "A combination of circumstances has shaped an unstable mood compounded by misinformation, mistrust, confusion, and fear," and there were "incipient signs of panic."[44]

The uncertainty the crisis had caused the American public to feel, of which the relentlessly elevating price of oil was both symptom and cause, seemed to corroborate the gloomy oil forecasts of both past and present. The Club of Rome, which had chastened the West for its unsustainable economics in 1972, appeared prescient. The authors of *Limits of Growth* had foreseen an era in which the price of nonrenewable resources would rise uncontrollably as supply ran out of control. Could it be that the world had already crossed the point of no return? E. F. Schumacher, author of the 1973 *Small Is Beautiful*, had warned the West that economic growth and industrial bigness could have dire consequences, fretting over the world's rising appetite for oil. The 1973 oil crisis seemed to signal the beginning of the end. The West had to adapt to a new energy reality, or perish.

Even the Shah of Iran believed that the oil industry's days were numbered. In a speech that was published in the Middle East Economic Survey at the close of 1973, the Shah projected that oil as an energy source "is going to finish" in "30 years' time." The old order of energy—sufficient oil at a low price—was over forever. That was no real loss, said the Shah—far better to use oil for the "noble purpose of the petrochemical industry or the medical industry" than to burn it as fuel. As for American nostalgia for the pre-1973 oil order, it was of little avail. Those in the "industrialized world" will "have to realize that the era of their terrific progress and even more terrific income and wealth based on cheap oil is finished. They will have to find new sources of energy. Eventually they will have to tighten their belts."[45] There was no mistaking Tehran's message: the good old days are over, and you had better get used to it.

Opinion polls around the time of the embargo, which had not bothered much with gauging public views about oil prior to the 1960s, underlined how pessimistic Americans were about their energy predicament. In a Harris survey from April 1973, before the embargo had even begun, more than half of respondents believed that "the U.S. was running out of oil and natural gas." Forty-seven percent of respondents described the 1973–74 "energy crisis" as "very serious." Another 30 percent called it "somewhat serious." Before 1974, the public had never ranked energy among its list of "important problem[s] facing the nation," according to Gallup. That year, however, it skyrocketed to the top of the list, only to decline once the crisis ended. When asked why they believed the crisis was so serious, a majority (75 percent) of Americans polled said it was that they expected "sharp price rises in the cost of gasoline and heating and air conditioning." Only half of those surveyed believed that "U.S. know-how is so good we'll find enough energy to meet our needs without a lot of trouble." Others apparently believed that permanently higher prices might be in store.

Americans were deeply cynical about the reasons for the crisis, and what they were hearing from official channels did not help. They tended to judge the hype as manufactured by the oil industry in order to make a quick buck rather than real problems with supply and demand. As one analyst describes it, "The public seemed to be split into two groups; those who felt the crisis was genuine and likely to worsen, versus those who made up the largest group, who thought it was contrived." A CBS/*New York Times* poll conducted in February 1974 revealed that nearly three-quarters of Americans did not believe that the energy shortage was "for real," but thought instead they were "just being told" that by elites. When asked who was to blame, oil companies were far and away the most common reply (56 percent), with the president the second most culpable (39 percent). OPEC and Arab leaders were a distant fourth (22 percent). In an atmosphere of such suspicion, those who tried to explain what was going on, whether newscasters, oil industry spokespeople, or politicians, would not have gotten far in dampening any anxiety about the state of the oil market.

Oil prices stopped climbing after the embargo passed, but they did not fall either over the next few years, which kept U.S. officials and the rest of the country on tenterhooks. A barrel of oil imported to the United States cost $12.52 in 1974, on average. It crept to $13 the following year, then just above $14 by 1978. Given the rampant inflation at the time, such stagnant oil prices actually meant that annual real oil prices declined slightly between 1974 and 1978, though consumers would not likely have paid much attention to that. The stratospheric jump in prices between 1973 and 1974 did not wear off quickly; things did not simply get back to normal in the oil market. A collective sigh of relief from drivers had come when the embargo ended, but the preoccupation with what OPEC might do next did not end. The number of articles on energy issues in *Time* and *Newsweek*, two of the country's most beloved magazines, in 1975 was nearly as high as it had been the year prior, in the immediate wake of the crisis. The themes of the pieces, according to one study, were "oil politics" and "OPEC politics"—two topics that had not ever stayed on the radar before the uproar in 1973.[46] The new normal in the oil market—of prices roughly four times what they had been in the decade prior—had settled in, and with it the attention of American consumers and their elected leaders, however reluctantly.

"Higher Prices Stretching out at Least over the Next Decade, and Likely Longer"

The volatility of 1973 had spawned a proliferation of what had been a relatively rare thing in the oil industry prior to that time: price forecasts. "Everybody's doing it," wrote *Oil and Gas Trends*, a monthly industry newsletter, summed up the growing cottage industry. "There are now so many energy forecasts around that people have started compiling opinion polls based on them." Whether all these predictions were of any use was the real question. "The distinctive thing about opinion polls," the industry reporter opined, "is that they tell you quite a lot about people's opinions and blessed little about the subject of their opining."[47] The IEA's comparison pulled together an astounding corpus of nearly 80 forecasts published between January 1977 and June 1978. Among the forecasters were the IEA, foundations (The Petroleum Research Foundation, The Rockefeller Foundation, et al.), corporations (Exxon, Shell, Sun Oil, et al.), U.S. government bodies (the CIA, Department of the Interior, et al.), foreign government bodies (Japan's Ministry of Trade and Industry, Canada's National Energy Board, et al.), and countless others. The reason for the explosion of forecasts was simple: the price of oil, a number of major economic significance to businesses, governments, and consumers all over the world, was making bigger moves than most anyone could remember. The fate of nations and economies, not to mention political careers, hung in the balance. Those willing to pay for price forecasts wanted to know whether another 1973 was around the corner. Yet predicting where the market was headed a decade or two in the future, even in the age of computers, would remain as much art as science, determined as much by subjective judgments and assumptions as by the relatively few certainties associated with future supply and demand.

These forecasts all displayed tremendous unanimity, despite disagreement over the particulars, in their essence. Production growth in places like Alaska and the North Sea, they agreed, might be enough to push the inevitably day of reckoning beyond the early or mid-1980s, but they were hardly large enough to solve the longer term problem. They were a temporary pain killer, not a true remedy.[48] The forecasts shared the assumption that oil demand worldwide, which had been growing at about 4 percent per annum between 1976 and 1978, would grow at that quick pace for the next decade or two. Such a trajectory would have indeed created an extraordinarily tight market, had it occurred (which it did not) and production figures proceeded as estimated (which they did not). Most forecasters saw another oil crisis occurring "within another decade or so hence, in the second half of the 1980s, when demand would once again be at the very edge of available supply." Even if physical supplies were sufficient, they believed, uncooperative sovereigns would pinch production, leaving the market thirsting for more and prices extraordinarily high and rising. This consensus view—of rising oil prices for which there was no obvious escape—was more than just a reflection of the times. It was a paradigm that would help shape the market over the next three years.

Several of the forecasts stood out. The CIA, in its first unclassified energy outlook, saw bad news piled high over the entire length of its forecast to 1985. Washington's top spy agency believed that the world's taste for oil would soon overwhelm what it could produce. Although that notion was all-too-common among oil analysts in 1977—the agency conceded that its forecast "broadly resembles other official and private forecasts"—it differed from the pack in that the CIA was "more pessimistic about the implication." OPEC could not keep up its current pace, the Agency argued. Saudi Arabia would need to produce "between 19 million and 23 million barrels per day if demand is to be met in 1985" (a full five times more than what it actually would, and around double what it ever has since that time). The Soviet Union's oil production had just begun to enter an irrevocable decline, or soon would. "In these circumstances," the agency declared, "prices will rise sharply to ration available supplies no matter what Saudi Arabia does." Unless people started using less oil, the market's fate was more or less sealed: higher prices stretching out at least over the next decade, and likely longer.[49] As if the forecast itself was not scary enough, coming from some of the nation's most terrifying and well-regarded strategists, as part of an objective reading of the tea leaves, it was all the more ominous.

Other voices, official and unofficial, would also chime in. Six months later, in October 1977, U.S. Secretary of Energy James Schlesinger put the warning in even more dire terms. A "major economic and political crisis in the mid-1980s" was on its way, "as the world's oil wells start to run dry and a physical scramble for energy develops." The ensuing crisis, brought about by sky-high oil prices and absolute shortages, could bring the United States "a degree of political and social unrest of the kind we did not see in the 1930s" during the depths of the Great Depression.[50] The Rockefeller Foundation also issued a warning to the world about the dire future of oil supply. The end of sufficient supply was nigh. "The weight of expert opinion," the report stated, "reflected in numerous studies based on the quantitative analyses, is in substantial agreement on one key point: the world is presently heading toward a chronic tightness, or even severe shortage, of oil supply," which would strike during the late 1980s. By that time, "concerns about oil will have multiplied. The present surplus will have faded away," leading to "competition among governments for supply [that] could be intense and unprecedented." The Rockefeller Foundation report continued: "The world community could be confronted by an unprecedented situation. Available oil supplies will be insufficient to sustain economic growth and achieve political and social objectives or even, perhaps, to maintain existing standards of living." Governments worldwide had to intervene; the "price of failure" was too dire to contemplate, with no less than world peace hanging in the balance.[51]

President Carter was sympathetic to this view of the oil supply crisis facing the world. It was a pessimism that would come to define the Carter presidency: his televised expressions of worry over a world running out of oil—not just in the United States, but worldwide—among the most indelible images of his time in office. As one historian has described it:

> What gripped Carter's imagination was the dreaded pincer movement of rising U.S. demand for oil and declining domestic production. The inevitable result would be rapidly increasing imports, made all the more dangerous because it spelled growing dependence on the Middle East. This scenario was set in context of a world supposedly running out of oil, even eventually in the Middle East.[52]

Although disturbing and fearful, the essence of the president's paradigm—that the United States needed to wean itself off foreign oil, lest it face rising oil prices and economic vulnerability through the end of the twentieth century—was totally in line with what many Americans, including energy experts, believed lay in store for the country.

"Likely to Get Progressively Worse through the Rest of This Century"

On April 18, 1977, Carter aired his now famous "unpleasant talk" with the American people, a nationally televised address in which he set out just how bad the energy crisis the country was facing really was. For Carter, it was "a problem unprecedented in our history":

> With the exception of preventing war, this is the greatest challenge our country will face in our lifetimes. The energy crisis has not yet overwhelmed us, but it will if we do not act quickly. It is a problem we will not solve in the next few years, and it is likely to get progressively worse through the rest of this century.

For those Americans who "doubt that we face real energy shortages," he informed them that simple arithmetic undeniably proved otherwise:

> The world now uses about 60 million barrels of oil a day and demand increases each year about five percent. This means that just to stay even we need the production of a new Texas every year, an Alaskan North Slope every nine months, or a new Saudi Arabia every three years. Obviously, this cannot continue.

The bottom line was clear: "we could use up all the proven reserves of oil in the entire world by the end of the next decade." Without "profound changes" in how people used energy, "we now believe that early in the 1980s the world will be demanding more oil that it can produce." The effects would be much more profound than just higher prices. Even they would not solve the total shortage that threatened the country. "Within ten years," he proclaimed, the country "would not be able to import enough oil—from any country, at any acceptable price." The result would be an "economic, social, and political crisis that will threaten our free institutions."[53] It was a dark speech, and yet Carter was hardly alone in worrying over the near future of American energy.

"Panic in a Chaotic Market Swept Everything before It"

If the rhetoric from the Oval Office had primed the American mind for an oil crisis, the political turmoil of 1979 and 1980 set a match to the gunpowder. This time, the trigger would be the Islamic Revolution in Iran, which started in late 1978. It would mark an end to the fifty-year reign of the Pahlavi dynasty. The revolution culminated in a religious scholar from central Iran, Ayatollah Khomeini, declaring the country an Islamic republic, introducing Islamic law, and assuming leadership of the country for life. The Shah's fall immediately caused Western oil firms to wonder what might be in store for them in a country that, in late 1978, accounted for 10 percent of all world oil production, surpassed in output only by the United States, the USSR, and Saudi Arabia. No sooner had the Shah fled the country in January 1979 than Iranian oil production crashed to a mere 40,000 barrels per day, as Western oil companies departed en masse. Production would bounce back to 4 million barrels per day by April, but the short-term damage of such a massive amount of oil taken off the market so suddenly was catastrophic. The price of oil imported to the United States doubled over the course of 1980 under the strain, rising from around $15 to $30 per barrel.[54]

The 1979 oil crisis passed through two phases, according to several historians.[55] The first phase lasted from December 1978 until the autumn of 1979. During this phase, the drastic reduction in Iranian production caused exports to collapse for several months. Saudi Arabia stepped in to fill the void, ramping up its production, although it did not fully make up the difference. Global production fell by 2 million barrels per day between the last quarter of 1978 and the first quarter of 1979. Panic buying ensued as downstream operators hurried to put crude and products into storage. This precautionary pressure on supplies put a heady premium onto any oil readily available on the spot market, continuing until commercial storage facilities all but ran out of room to store more oil.

The second phase of the crisis began in the autumn of 1979. No sooner had spot prices for oil began to ease than war broke out between Iran and Iraq in September 1980, which pushed average U.S. imported oil prices from $34 per barrel to $39 by February 1981. The war cut into OPEC production, which fell by 4 million barrels per day between its 1979 peak and August 1980. Mayhem in both countries saw Iraqi oil production fall from 3.5 million barrels a day to less than 1 million per day just months after the outbreak of hostilities. Iran's output, which the revolution had already decimated, declined only slightly over that time. Again, other OPEC member countries were able to partially offset the lack of production from the two countries, but only at a lag, though global output did recover over the course of 1981. The second phase of the crisis was over by the end of 1980.

What made the second oil crisis so devastating for the oil market? With global oil consumption on the rise for several years, the market had been preparing for tight conditions in light of what appeared to be unyielding demand growth into the future. Expectations of higher prices and continued tightness in the market in the future magnified panic buying. Contractual supply channels the world over had been destroyed by the Iranian revolution. The severing of Iran's production ties to the rest of the industry, if only short-lived, caused a domino effect of disruption among the complex ties of suppliers, producers, refiners, and transporters, among others, that existed in the gigantic world oil trade. The seamlessly integrated world of the Seven Sisters, broken by the wave of nationalizations in the 1970s, was now in complete tatters. More nationalizations in 1979 and 1980 meant the oil majors' share of trade volume fell by 10 percent between 1978 and the end of 1979.[56] Aware of this temporary situation, oil exporters exploited the upheaval to reap windfall profits. Opportunistic behavior on their part moved prices higher, and oil companies were not above using the crisis where they could to turn an extra profit.

Yet the emotions of oil market participants, and the fog of war in which they operated, multiplied the price drama. A lack of credible information about what was going on in the market increased the chaos and exaggerated the whipsaw of prices. The reality of the crisis was not nearly as bad as it seemed in the moment. Yet it appeared as though this crisis was the ugly realization of the previous years' gloomy forecasts. It was as if the long-term supply crunch of which some analysts had foretold had finally arrived, bringing with it an era of much higher oil prices with no end in sight.[57] Panic amidst chaos, exacerbated by fears of future supply shortfalls—as old as the oil industry itself, though never on such a scale—had moved prices much more dramatically than the actual loss in supply really merited. As Parra puts it:

> [No physical factors] explain the tripling of oil prices that took place within a period of months in the absence of any shortage of crude. The economic disaster that took place in late 1978 was a triumph of mind over matter, faith over fact and belief over the drumbeat of contrary evidence. Panic in a chaotic market swept everything before it, consumers, producers, companies, governments; but there was never a shortage of crude.[58]

Stockpiling took place on a tremendous scale, both among retail and institutional customers. The worst-case scenarios of the forecasters played themselves out in real time by means of a full-blown oil market panic. It was a run on the bank, so to speak—but instead of withdrawing cash, buyers withdrew oil. Consumers all along the supply chain, believing higher prices and/or more shortage awaited them, bought as much oil as they could get their hands on. That way—should prices go up as they expected them to—they could dump the barrels back on the spot market for a terrific profit. Motorists, industrial users, utilities, and refiners were all to blame. According to one estimate, the rush by these consumers to build emergency inventories and store gasoline alone took a massive 3 million barrels per day off the market.[59]

"The Inevitable Movement of Prices ... Will Be Upward"

But such nuances were mostly lost on Washington. The oil crisis left little doubt among the leading lights of the Carter administration that a new era of scarcity had arrived, itself symptomatic of a broader American decline—and at the hands of "a few desert states," in Carter's words, adding insult to injury.[60] James Schlesinger, departing from his post as U.S. Secretary of Energy in the summer of 1979, described the trouble in the oil market in the starkest possible terms. "Today we face a world crisis of vaster dimensions than Churchill described half a century ago—made more ominous by the problems of oil." In the decades ahead, he warned, the conflict among nations for "control of the oil tap" could become the "decisive element in the East-West struggle." For the United States and other consumers, it was a battle that was almost certain to be a losing one. "There is little, if any, relief in prospect," he prophesied. He saved his harshest scorn for those of his fellow Americans who believed that the United States might soon find a way to produce more oil from hitherto-undiscovered resources:

> The widespread fantasy that the United States contains enormous proven reserves of oil, which are hidden and unreported, is something difficult to explain. It may simply fall under the heading of psychopathology. But it is grist for the mill of those unscrupulous few who would ... suggest in various ways that somewhere out there, there exists an enormous cache of oil that will wash away all other discomforts. Demagoguery is unlikely to provide an effective substitute for crude oil.

For the oil market, an era of more or less permanently high prices was in store. The "inevitable movement of prices, save in the face of worldwide recession, will be upward," he predicted, reflecting the "increasing pressures of demand against constrained supply." In short, as far as the American people were concerned, "The energy future is bleak and is likely to grow bleaker in the decade ahead."[61] Schlesinger reiterated his alarming message to the *Wall Street Journal* in August 1979. "Quite bluntly, unless we achieve greater use of coal and nuclear power over the decade, this society just may not make it."[62]

Two of the world's most powerful oil companies also saw serious trouble ahead when it came to getting oil out of the ground. A 1979 British Petroleum policy brief painted a picture of the supply problem that was not dissimilar from the peak oil school that would become popular several decades later: "For a decade or more, there have been signs that the world's oil resources were being depleted at a rate that could not be sustained." No, the world was not "going to 'run out' of oil—as the media often put it." But it would have to brace itself "to a supply which *declines.*" Global oil production would "almost certainly" begin to shrink "in the next five years." Oil production outside the Soviet bloc would peak no later than 1985, and all told, world oil production would be 25 percent lower by the end of the century than it had been in 1979.[63] Exxon's 1979 annual report, which came out in the spring of 1980, foresaw "the balance between world energy demand and available supplies" being "precarious" in the following years. The old era of cheap oil was over. Yes, "significant oil and gas reserves will be found" during the 1980s, but "because the world has been using more oil than it has found, much of what is discovered will merely help offset declining production from existing fields." Going forward, "virtually all new petroleum reserves" would be "expensive to find and develop." The challenge, then, would be to try to thrive in an era of potentially declining supplies and costly production.[64]

The CIA, buoyed by the apparent prescience of its predictions of market panic made two years earlier, decided to double down. "The gas lines and rapid increases in oil prices during the first half of 1979 are but symptoms of the underlying oil supply problem," the spy agency's top thinkers wrote in August 1979. "The world can no longer count on increases in oil production to meet its energy needs." Geology was having its revenge. Oil was being consumed more quickly than it was being discovered, which meant that output would reach an insurmountable peak no later than a decade hence. "The predominant view among geologists" was that "the chances of discovering enough quickly exploitable oil to offset declines in known fields are slim." This was Malthus writ large: "In its broadest scope," the CIA declared, "the world energy problem reflects the limited nature of world oil resources." While OPEC producers could in theory decide to try to increase their production capacity, the CIA saw little reason why they would do so, even if they were able to. More likely they would begin to produce less, not more, to suit their preference for high prices. In terms of supplies, world oil production "probably will begin to decline in the mid-1980s." The United States and the rest of the developed world would simply have to "adjust to a slow growth and a stable or declining oil supply" over the long term, which will "impact . . . sharply" the global economy. What did all of this mean for the market? Real oil prices were likely to rise again and again "in spurts." Temporarily weak demand gave the "illusion of ample oil supplies, masking once again" the inevitable long-term crisis to come.[65]

The agency made another yet more alarming set of predictions, which came courtesy of testimony by the director of the CIA, Admiral Stansfield Turner, published in May 1980. We were right, Turner trumpeted—and where we were wrong, it was because we were too *sanguine* about the future of oil. In his words:

> In 1977, the Agency produced its first unclassified report of the inter-national energy outlook. At the time, the report was roundly criticized as being overly pessimistic. As things turned out we were not pessimistic enough. The events of last year have once again demonstrated that the energy problem is a serious one and that it is with us now.

Oil was about to hit a wall—or a ceiling, at least. "We believe that world oil production is probably at or near its peak and will decline throughout the 1980s." There was not enough of the black gold left to go around. "Simply put, the expected decline in oil production is the result of a rapid exhaustion" of "conventional crude oil." Yes, huge amounts of tar sands in western Canada and other sources of liquid oil that were extremely hard and expensive to ex-tract were known to exist, but all the tar sands in the world could not overcome the fact that the easy stuff was nearly gone. "There is good reason to believe," the CIA director went on, "that the most prolific oil producing areas have al-ready been located and drilled. Even with modern technology, the chance of finding new giant fields is diminishing." As a result, "neither efforts to find more oil or to provide fuel substitutes can compensate for the shortfall during the next ten years." Would this new era of scarcity, which the oil crises of the 1970s had already ushered in, lead to the outbreak of war? Undoubtedly, the admiral opined. "Politically, the cardinal issue is how vicious the struggle for energy supplies will become."[66]

Many in Washington's halls of power viewed the events of 1979-80 as the first stage in a protracted drama of shortage and violence, rather than merely a discrete episode of panic. James Akins, the erstwhile voice in the wilderness who had risen to become U.S. ambassador to Saudi Arabia, raised his voice once again in 1979 over the crisis that was to come. Soon, he predicted, the world would "enter a period of permanent oil shortage" and "semi-anarchy in energy, with the richest and strongest of the consumers making bilateral deals with OPEC."[67] Henry Kissinger, commenting on the situation three and a half years after departing as Secretary of State, did not believe havoc in the oil market would be over any time soon. "All the preconditions ... exist," he would write on July 31, 1980, for "the next four to five years" to be a "period of crisis."[68] If a staff report prepared for the U.S. Senate's Committee on Energy and Nat-ural Resources in November 1980 was any indication, neither did many in Congress, in either party. Three months into the Iran-Iraq War, the future of energy geopolitics from Washington's perspective was looking increasingly conflict-prone:

The world will witness a growing struggle for secure access to oil through the end of this century and into the next. This gathering energy crisis deserves the highest priority in the councils of government. Few other problems are more complicated; few other problems will be more difficult to solve. Moreover, many of the policies which we are currently pursuing to deal with the energy crisis are only making it worse.

The report did not exactly strike a hopeful note. The committee called on the country to "reorient its approach to the gathering energy crisis." It was neither something which could be put off, nor would it disappear anytime soon, even with deliberate action. "We must recognize that the energy problem is not only a long-term issue but that it is an immediate concern." The report concluded.[69]

Senior officials in the United States, Japan, and Europe were not the only ones concerned about a worsening oil shortage. OPEC heavyweights harbored private doubts about their own wells. Francisco Parra, a former secretary general to the cartel, later wrote that "OPEC's major fear was one of premature depletion, something that has been endemic to the industry since its early days," and which the "extravagant development plans" of the early 1970s had "powerfully aroused." Rumors of shortage were in the air, and OPEC was not immune to them. "To be fully appreciated," Parra went on, the fear "must be seen against the background of an almost universal belief in industry and consuming government circles that energy would soon be in permanently short supply." Middle Eastern producers worried that should they open the faucet too much, they might hasten their own demise by selling their gold too fast, and for too little. The oil kingdoms also feared military action against them by the West should they prove uncooperative, and yet they bristled at the thought of being forced to produce beyond their revenue requirements. Doing so would open the door to a fate in which "after a period of disorderly development, they would be left depleted and cast aside."[70]

Saudi Arabia, despite its massive clout as a producer and reserve holder, was not above such concerns, according to Parra. A 1979 U.S. Senate Committee on Foreign Relations report, *The Future of Saudi Arabian Oil Production*, expressed skepticism about whether the Saudis could continue their production output. It had given Riyadh pause, and the Saudis were apparently not the only ones who feared the possibility of exhausting their reserves prematurely. Reportedly, "a group of Arab officials from Kuwait undertook a special visit to the ghost mining towns of the western United States to see their future for themselves. They returned, not surprisingly, shaken." In Parra's judgment, OPEC's steadfast refusal to lower posted prices and remove production ceilings during the crisis owed to a private fear on their part about how long their reserves would actually hold up. They finally agreed to raise the production ceiling to 9.5 million barrels per day in the third quarter of 1979.

From Gloom to Glut

As it turned out, the only specter with which OPEC would have to do battle as the 1980s dawned was not scarcity—but overwhelming, even problematic, abundance. Consumption was slipping. In 1979, the world had gulped down just over 65 million barrels a day of the stuff. But with oil prices elevated and the economies of the United States and Europe still reeling from the sharp jump in prices, consumers' habits began to change, causing oil consumption to decline sharply to just 60 million barrels per day in 1981. Why the pull back? For one thing, energy conservation efforts, which the world's leading economies had begun putting in place in the mid-1980s, were starting to bear fruit. The Energy Policy and Conservation Act, passed in 1975, had mandated that the average fuel efficiency of American-made passenger vehicles double within ten years. U.S. drivers were also spending less time on the road. Nationally, three years would pass before U.S. drivers began to rack up as many miles traveled as they did in 1979. Demand for oil also decreased as people found ways to generate power using energy sources less expensive than oil. Spurred by a wave of deregulations that allowed oil and gas prices to better respond to market forces, natural gas, coal, and nuclear power all began playing a larger role in the nation's electricity mix.[71] All of these factors added up to a declining global appetite for oil, in stark contrast to the never-ending upward march that nearly all predictions of impending energy doom had foreseen.

What was more, oil produced beyond OPEC's borders was flowing into the global market at a record pace, beating forecast after forecast, from a variety of countries. As usual, the prospect of huge financial rewards from record prices at the wellhead tempted the industry into remote regions around the world, and advances in exploration and production technologies provided the capabilities necessary to do what years earlier would have been impossible. Soaring prices did not work their magic on supply all at once, however. It would take years for the oil they called forth to come to market in scale, but eventually it did come.

A rising tide of oil from Alaska, Mexico, the North Sea, and the Soviet Union all lifted global supplies. It would take only three years from the initial development of the gigantic Prudhoe Bay on Alaska's North Slope, the largest oil field in North America, in the spring of 1977 for it to spew more than 1.5 million barrels per day of oil, amounting to nearly 20 percent of all U.S. crude oil production. Mexico was another success. The country had once been a mighty producer, accounting for 21 percent of all the oil production in the world in 1921, second only to the United States.[72] But public management, catalyzed by the 1938 nationalization of the country's oilfields, would dampen volumes for decades. Thanks to the discovery of sizeable fields in southern Mexico and Mexico City's urgent need for greater revenues, which led to greater investment in Pemex, the national oil company, production began to roar. It doubled between 1973 and 1977, only to double again by 1980, hitting 3.1 million barrels per day in 1981. The first major find in the North Sea, where the United Kingdom, Norway, and Denmark would each lay claim to a portion of the reserves, took place in 1969, the first oil reaching Cruden Bay by pipeline in 1975. As elsewhere, the price shock caused by the first oil crisis made pumping North Sea oil economically viable. Practically no oil was being produced in the North Sea in 1974. A decade later, though, it was putting out over 3 million barrels per day.[73] Meanwhile, the oil output of the Soviet Union, which CIA analysts and countless others had written off as doomed to dwindle, did not succumb. Instead, it swelled, rising from 8.6 million barrels per day to 12.1 million barrels per day in 1981, an astounding increase of 40 percent since the first oil crisis.

With consumption declining and new sources of supply springing up around the world, OPEC was beginning to feel the heat. To stabilize prices, it had to take action, but restraining its production in a unified fashion was easier said than done. Early in 1982, for the first time, it forged an agreement among its member countries to adhere to a strict production limit of 17.5 million barrels per day. No more producing beyond the agreed-upon quotas, the Saudis ordered, setting the ceiling at 50 percent of 1979 production levels. But Riyadh's OPEC brethren were less than amenable to such top-down austerity. Tehran, locked as it was in the midst of a brutal war with Iraq, bridled at the Saudis dictating how and when they could sell their national treasure. As OPEC continued selling, unswayed by the kingdom's entreaties, spot prices kept swooning. The market was rapidly moving from scarcity to glut. As inventories rose worldwide, OPEC, backed into a corner, took the drastic step of making its first major price cut. The Saudis, having had to accept a 34 percent drop in output by between 1981 and 1982 in an attempt to firm up the spot market, cut their posted price for Arabian Light crude by a full $5 to $29 per barrel. OPEC slashed its official prices by 14 percent between January 1982 and March 1983.

OPEC struggled from 1983 through 1985 to hold back the onrush of oil to a market that simply could not bear supply in such quantities. Oil pouring out of the British portion of the North Sea fields alone—no more than a rounding error a few years prior—had overtaken that of Algeria, Libya, and Nigeria combined. Saudi impatience with rampant overproduction among its OPEC peers, in the midst of such price calamity, was becoming palpable. When

OPEC ministers gathered in Taif, Saudi Arabia, in June 1985, Sheikh Ahmed Yamani, Riyadh's outspoken Minister of Petroleum and Mining Resources for more than two decades, read aloud a letter from King Fahd. The monarch roundly criticized those countries whose cheating on quotas and price slashing had caused a "loss of markets for Saudi Arabia." The kingdom would not countenance such recklessness forever. "If Member countries feel they have a free hand to act," Fahd wrote, "then all should enjoy this situation and Saudi Arabia would certainly secure its own interests."[74] There was little doubt that Riyadh was doing all it could to put the market on firmer footing, having ratcheted back its output to a paltry 2.2 million barrels per day by that time, only a shadow of what it had been producing five years earlier. Still more insulting, it was less than what the startup Brits were pumping in the frigid waters of the North Sea. In the weeks after huddling in Taif, however, the rest of OPEC showed no intention of heeding Riyadh's pleas.

As summer waned, the Saudis decided they had had enough. Yamani announced at the Oxford Energy Seminar in 1985 that the kingdom would no longer shoulder the burden of price stability through restricting its own share of the pie. He made it clear that the Kingdom would defend higher prices over the long term: "I have no doubt in my mind that there will be a shortage in the supply of oil" in the decades ahead, he revealed, a prediction the Saudis were uniquely able to bring to pass if they desired. But the days ahead were "vague and unknown." As Yamani made clear, Riyadh was contemplating taking a step that had never been done before: intentionally flooding the market. "Most of the OPEC member countries depend on Saudi Arabia to carry the burden and protect the price of oil. Now the situation has changed," he declared. "Saudi Arabia is no longer willing or able to take that heavy burden and duty, and therefore it cannot be taken for granted. And therefore I do not think that OPEC as a whole will be able to protect the price of oil."[75] He was right. It was not long before OPEC's 800-pound gorilla began opening the spigots and changed the way they priced their oil, adopting so-called netback pricing, allowing the kingdom to undercut official rates elsewhere.

The result was a near-complete reversal of the conditions that had defined the market since the early 1970s. There was indeed an oil crisis unfolding— but exactly the opposite of the one that the United States had come to fear. This time, it was oil producers, not consumers, who woke up to a new world. This was a crisis of previously unimaginable abundance, rather than scarcity. Although not by a large margin, supplies were coming onto the market at an unsustainable pace, forcing prices downward. Saudi oil had finally broken through an already leaky dike. Exporters, not importers, were now the ones in the precarious position of worrying over with whom to place their oil and how badly their revenues would be squeezed. As prices spiraled downward, buyers were locked in a race to the bottom, trying to sell their oil at whatever price the market would bear. Subtle gestures by some OPEC producers to offer a bargain here or there could not contain the chaos. Sovereign sellers tried to structure contracts in innovative ways to lock in market share at a price they could countenance. As to their goal of preventing prices from falling, though, the market had other ideas.

The severity of the crash was stunning. Between the fall of 1985 and the following summer, the price of the United States' benchmark crude oil, West Texas Intermediate, collapsed to just one-third of its former value, dropping to just $9.83 per barrel. Dubai crude struck bottom at $7 per barrel.[76] By September 1986, prices had risen to $14.36, only to drop back to $13.33 in November.[77] The oil crises of Nixon and Carter unraveled, as the cost of crude oil in the United States, once adjusted for inflation, hit levels in 1986 they had not lingered at since the distant, Edenic days before the embargo. American drivers breathed a sigh of relief as real gasoline prices fell to their lowest levels in a generation, less than $1 per gallon in nominal terms. The tables had finally turned. Now it was OPEC's turn, not the United States and other major importers, to worry about what it meant for their coffers. OPEC revenue in 1986 was less than a quarter of what it had been at the start of the decade, when the figure had reached a record $275 billion.[78] Unfortunately for OPEC, this time there was no John D. Rockefeller, Texas Railroad Commission, or Seven Sisters to put things back together again. Gasoline prices in the United States, adjusted for inflation, would not return to the lofty levels of 1985 for another two decades.

Oil was everywhere, but predictions of shortage, prolific just a couple of years earlier, were nowhere to be found. Their well had run dry—at least for the time being.

5

"A Permanent Radical Rise in Oil Prices"
Peak Oil Descends on Wall Street,
1998–2013

"Oil shocked." It was a fitting headline for an *Economist* article describing the state of the oil market in 1998.[1] Shock was indeed what the market was experiencing. But this was hardly the type of shock that Americans who had lived through the chaotic leaps in oil prices during the 1970s were accustomed to. It was the opposite extreme: oil had fallen to radically low prices and was dipping lower by the week as the year came to a close. A barrel of West Texas Intermediate was selling for around $18 on New Year's Day, 1998. Eleven months later, it had collapsed to just more than half that. A gallon of gasoline cost less than $1 a gallon in the United States. Oil had not been this cheap since the summer of 1986, after Saudi Arabia had decided it was tired of trying to maintain high prices amid a lack of support from its OPEC brethren. Adjusted for inflation, the price of oil had fallen to a twenty-five-year low.

"If You Are Having Haunting Visions of Long Lines and $2.50-per-gal. Gasoline, Relax"

A number of forces combined to push prices downward. The East Asian financial crisis was the primary culprit. Precipitated by a collapse of the Thai baht in the summer of 1997, the panic saw the region's stock markets fall by as much as 60 percent and caused oil demand in that part of the world, a pillar of global demand, to pull back sharply. It also meant that demand for oil elsewhere grew more slowly than usual. Even as demand worldwide was wilting, oil continued to flow onto the market, unrestrained by OPEC. Iraqi oil had begun to surge onto the global market for the first time since the Gulf War, rising from just shy of 600 barrels per day in 1996 to nearly four times that amount two years later.[2]

OPEC, caught flat-footed by the Asian crisis, was in disarray. In November 1997, just as oil prices were starting to sink, OPEC ministers had agreed at one of their regular meetings to raise their production quota by 2 million barrels per day. They did so on the mistaken belief that world consumption would continue to increase at the same rate in 1998 that it had between 1996 and 1997, during the heyday of the Asian economic miracle. Raising quotas was an easy way to bring them closer in line with actual production numbers, since many of the exporters were cheating, producing at their maximum capacity regardless of the rate OPEC allotted them. As best as the OPEC ministers could tell, the fundamentals of the market appeared to justify the move. As it turned out, their timing could not have been worse. OPEC had begun to ramp up production precisely as demand was starting to fall. OPEC met several times in 1998 in a vain effort to establish production quotas severe yet realistic enough to stop prices from sliding. Some members of OPEC, most vociferously Venezuela, were loath to cut back their output, particularly frustrating Saudi Arabia, which viewed Caracas and others as stealing its rightful share of the market. Amid the squabbling, prices slipped to $10 per barrel, with some grades selling for as low as $6, by the end of 1998.[3]

It took a 40 percent drop in prices between October 1997 and March 1998, which sliced billions out of OPEC revenues, before Saudi Arabia was able to orchestrate an emergency agreement to try to rein in production from other OPEC producers, who were loath to lose market share to newly resurgent Iraqi oil exports. Moreover, the crisis had caused investors to pile into safe haven assets, such as U.S. treasury bills, stocks, and other U.S. dollar-denominated assets, which pushed the value of the dollar up by nearly 20 percent in the six months after the crisis began.[4] Because crude oil is priced in U.S. dollars, the sharp jump in the currency's worth, relative to other currencies, meant that oil in the United States was significantly cheaper. It also meant that oil was more expensive in many hobbled emerging-market currencies, which exacerbated a decline in demand in those parts of the world. Adding to the demand downturn was a Northern Hemisphere winter that was one of the mildest on record, limiting heating needs. In the meantime, the amount of oil in storage around the world surged going into the winter of 1998, the visible evidence of a glutted market.

Oil was cheap, but whether that was a good or a bad thing depended on who you asked. For American drivers, it was Eden. Car buyers embraced trucks and sport utility vehicles over smaller cars, eschewing fuel-efficiency in favor of girth and brawn. Why economize on gasoline, consumers reasoned, when low prices allowed you to get more car and more miles traveled for the same outlay at the pump? "What sells these days are the biggest things in the showroom," one Florida newspaper trumpeted. A new wave of "rough tough land yachts" rolling off assembly lines across the country—the Chevrolet Suburban, the Lincoln Navigator, and the Ford Expedition, to name a few—could not be produced fast enough to meet demand.[5] "Bigger may or may not be better," in the words of the *New York Times*, but "automakers are scrambling to build the behemoths Americans will buy."[6]

While car buyers raved, oil producers and the companies that serviced them mourned. "Those of us in the oil patch have seen much better days," lamented an oil-drilling executive in Tulsa.[7] It was an understatement. Oil companies slashed jobs as well as capital expenditures on exploration and production as they struggled to stem the bleeding. Companies took their drilling rigs out of production in droves due to the collapsing value of their product. The number of active oil rigs drilling in the United States fell from 392 in September 1997 to just 111 by March 1999.

But even those measures were not enough. The finances of the world's major oil companies were in for a shakeup unseen since the U.S. Supreme Court had undone the Standard Oil behemoth a century earlier. With oil prices so low, even the industry's most powerful players were forced to go to extraordinary lengths. Coming to terms with dramatically altered market realities meant that companies had to find a way to boost efficiencies, contain costs, and leverage technology and human capital to stay in the game. All of these needs pointed toward one strategy, which oil executives one by one began to adopt: merge with or acquire a competitor. A wave of mergers swept over the industry between 1998 and 2002, making many of the erstwhile oil "majors" now the "supermajors." BP swallowed Amoco in 1998, the largest foreign takeover of a U.S. company. Exxon and Mobil merged in 1999 in an $80 billion deal. TotalFina took over Elf the same year. Chevron bought Texaco for $39 billion two years later, the same year that Conoco and Phillips announced their merger. A host of other companies would be touched by the merger mania over these years, some of them now just names from the past: ARCO, Mitsubishi Oil, YPF, Getty, Enterprise Oil, Nippon Oil Company, Amerada Hess, and many others.[8]

The glut had a way of eliciting rosy sentiments about a future of ultra-low oil prices from even the most tenured analysts, which suited American audiences just fine. Bearish market forces, rather than anticompetitive politics were now in the driver's seat, they argued. "The world is dramatically changed," observed Robin West, chairman of Petroleum Finance Corporation. The oil market "really is transparent and efficient."[9] Some predicted a realignment of the oil order, with Middle Eastern producers losing their exalted place in the energy firmament once and for all. "This may be the end of the old OPEC," opined the chairman of the Petroleum Industry Research Foundation, John Lichtblau.[10] The oil economy had been reborn, with consuming countries like the United States now in the position of privilege, with little threat to the high oil prices of decades past raining on the parade. "In the old days the question was low or high prices, and everything rode on those numbers," wrote one American energy expert. "Now the question is low or very low prices."[11] A futuristic array of new information technologies, the growing role of natural gas in the energy mix, and the diverse sources of oil around the world meant that the "industry has changed fundamentally since previous shocks that seemed to portend ever higher prices," he told *Time*. "If you are having haunting visions of long lines and $2.50-per-gal. gasoline, relax."[12]

Enter Peak Oil: "One of the Great Turning Points in History"

And yet, amid apparently limitless abundance, a small pocket of obscure analysts were not nearly so carefree. A few lone voices began to warn that the oil industry would soon bump up against unyielding geological constraints, with almost unthinkable implications for world oil prices. Writing in *Scientific American* in March 1998—the very time at which inflation-adjusted prices were the lowest they had been in twenty-five years—Colin Campbell, a geologist, and Jean Laherrère, a petroleum engineer, laid out an elaborate case for why the low prices the world had enjoyed up to that point in history were about to disappear for good. They acknowledged that oil analysts in the 1970s had claimed that the world was running out of oil, only to have to eat crow a decade later. But this time was different. "The next oil crunch will not be so temporary," they ominously warned.

Their bottom line was as simple as it was shocking: The world's oil production would begin to decline for good no later than 2010, and possibly as early as 2004. "Peak oil"—in other words, the maximum rate of oil production globally—was in sight.[13] After the world had hit peak oil, they argued, oil prices would then start to rise forever, "unless demand declines commensurately." Just twelve years in the future, they reasoned, the Middle East will have pumped more than 50 percent of its reserves. With that giant region running on the second half of the tank, the world's oil production would enter terminal decline. Global reserves would yield less and less of the stuff. "The world is not running out of oil . . . at least not yet," they wrote. There was still much more to pump. But oil had reached its tipping point. "The end of the abundant and cheap oil on which all industrial nations depend" had finally arrived. Once production begins to taper off, beginning in a decade or so, a new era of endless price hikes would ensue.

The two engineers based their case on several claims. There was reason to believe that OPEC countries, whose quota allotments were based on the size of their reserves, were overstating how much oil they were sitting on. Companies' estimates of their own booked reserves were also suspect, inflated to juice the price of their stock. They noted, correctly, that there is a tremendous amount of haziness and opaqueness when it comes to estimates of any given country's oil reserves, partly because definitions of reserves can vary from place to place, but also because estimating reserves is an "inexact science" and many countries are loath to reveal what they know about their own holdings and production capacity. Notwithstanding these analytical roadblocks, Campbell and Laherrère were confident that they knew exactly how much oil remained. The industry could extract "another 1,000 billion barrels of conventional oil," they concluded, a shade more than the 800 billion barrels it had already pulled out of the ground. Given their forecast for oil consumption, that meant that half the world's oil would be gone no later than the first decade of the twenty-first century. It was a truth of history, they declared, that "the rate at which any well—or any country" or the entire world, in this case, "can produce oil always rises to a maximum and then, when about half the oil is gone, begins falling gradually back to zero."[14] And just a few years hence, back to zero was exactly where the world would start heading.

Coming from a respected outlet like *Scientific American*, it was quite a startling prediction. The few others who had published similar outlooks in the late 1990s never got anywhere near the traction that Campbell and Laherrère did. John Edwards, a geoscientist at the University of Colorado, had written a piece in the summer of 1997 for the bulletin of the American Association of Geologists that put forward an almost identical argument. World crude oil production, he wrote, would max out at 90 million barrels per day by 2020 (it was around 64 million barrels at the time). The tipping point, at which time half the world's available oil had been pumped, would come in the year 2000. As a consequence, the world would then "face a shortfall in supply and a permanent increase in the price of oil." By 2050, he argued, no more than 18 million barrels of oil a day—roughly 1950 levels—could be pumped. It was a foreboding vision. "The next century will see an irreversible change in world energy supplies," he prophesied, "which will lead to alterations in the lifestyles of most of the world's population."[15] Richard Duncan, Walter Youngquist, Buzz Ivanhoe, and others all put forward similar cases, some with a more frightening tone than others, but all embracing a methodology identical to that of Campbell and Laherrère.[16] The "inevitable fuel crisis," an article published in the *Futurist* magazine in early 1997 warned, was a "grim fact" of the "permanent oil shock" soon to come.[17] All advised governments and concerned citizens the world over to make haste in preparing for its coming.

It was not the first time that Campbell, a Brit who had held senior positions at a number of oil companies, had called the end of oil. In 1991, he had published an obscure book arguing that the world's oil production would max out within ten years. The "epoch of declining production is about to begin," he warned. "Future generations will likely look back and see this inflection point . . . as one of the great turning points in history." He acknowledged how "staggering" it must be for his readers "to realize that we will have to face a radical new situation so soon: probably well before the end of the century, less than a decade away."[18] As the decade wore on, and oil supplies did not peak as he anticipated, he was forced to push the date of the still-imminent peak farther out into the future, just beyond the horizon. In a 1996 article he had cited "compelling evidence" that the "pending oil supply shortfall" would occur "possibly before 2000" (rather than the "no later than 2010" window he laid out in his 1998 prediction) rather than the following decade, as he would later argue. And what were the consequences of the upcoming energy crunch, as he foresaw them? Nothing short of a "permanent radical rise in oil prices." He acknowledged a "certain skepticism" among his opponents "because people have cried wolf before." But not to worry, Campbell assuaged: "This time the situation is totally different."[19] In time, even the doubters would see that "Petroleum Man will be virtually extinct this Century."[20]

Campbell's vision of an imminent oil collapse was exotic and alarming, but the methods on which he based his arguments were hardly new. They had their origins in the work of another petroleum geologist with a penchant for prophecy, M. King Hubbert. Half a century earlier, as a scientist at Shell's research labs in Houston, Hubbert had pioneered a simple statistical approach to estimating when oil production from a given region would reach its peak and enter into permanent decline. He observed that the life-cycle of many oil fields was not unlike a bell-shaped curve, rising and then falling in a neatly symmetrical way. Perhaps larger bodies of oil—say, all of the oil reserves in a given state, or even in an entire country—would also rise and fall in a similar fashion, he conjectured. If that were true, he reasoned, he could quantitatively estimate a decline curve for a country like the United States by treating the area as more or less a single oilfield, based on the size of the resource base and how fast it was being pumped.

In 1956, Hubbert used this approach to make a startling prediction, which he presented to a roomful of his peers at a technical conference. He had drawn by hand a roughly bell-shaped, symmetrical curve to match the path of oil production in the lower forty-eight states since the early days of industry to decades into the future.[21] Its implication was stunning: Oil production in the continental United States, he explained, would hit a peak between 1966 and 1971, only to decline sharply from then onwards.[22] The prediction was met with ridicule by nearly all his colleagues in the oil industry over the next decade. But the eccentric geophysicist from San Saba, Texas, would have the last laugh. In 1970, U.S. crude oil production outside Alaska and Hawaii topped out at 10 million barrels per day and began to fall. Hubbert, the controversial pessimist, was vindicated. To a rising generation of petroleum geologists, he was more than a rock star—he was a prophet.

Hubbert went on to predict the course of oil production globally. By his estimation, the total amount of oil in the Earth—the sum of everything that had been produced historically, known reserves, and projected discoveries—was 1.25 trillion barrels. He doubted in 1956 that all the world's oilfields could ever put out more than 12.5 billion barrels per year. If he was right, he argued, oil production globally would reach a ceiling in the year 2000. When Colin Campbell wrote in 1991 that world oil production would likely peak around the end of the century, he was merely picking up where Hubbert, by then deceased, had left off. But would the Hubbert curve prove as prescient this time as it had in 1956?

$30 Oil: "What's Going on Here?"

By the spring of 1999, oil prices had done an about face, reversing their decline during the bearish mayhem following the Asian financial crisis. The price of West Texas Intermediate (WTI) crude oil, a price that served as a world benchmark, bottomed out around $10 per barrel as 1998 turned into 1999. OPEC members met in Vienna in March to finalize a round of production cuts, forging an agreement that would take more than 2 million barrels of output per day off the market over the remainder of the year. Most of the officials had hammered out the broad details of the plan at a meeting in The Hague eleven days earlier. Even some non-OPEC members, like Mexico, took part in lowering production and exports. As the year wore on, it was clear the plan had done the trick. Asian demand was recovering briskly; robust economic growth in the United States was also adding fuel to the fire. Oil producers outside of OPEC, meanwhile, struggled to lift their output.

As global oil inventories declined, prices quickly rose to alarming levels. By August, spot WTI prices had doubled from their January lows, hitting $20 per barrel. When Iraq halted exports in November in a showdown over United Nations sanctions, just as the Northern Hemisphere was ramping up its demand for heating oil, the market leaped to $28 per barrel—the highest prices seen since the Gulf War a decade earlier. This was rarified territory. Over the last decade and a half, benchmark crude prices had oscillated neatly between $15 and $25 per barrel, except during the war over Kuwait. Now, in the space of a year, prices had nearly tripled, breaking through their long-standing trading range. It was enough to evoke unpleasant memories of the oil crises of the 1970s in the international media. A year earlier, it was the oil producers who were feeling the pain, but now consumers were worrying over what lay in store. Energy Secretary Bill Richardson, in a hurried attempt to calm the market in early December, told reporters that the Clinton administration would not shy away from intervening if these "dangerously high" oil prices continued to threaten the economy.[23] Despite the White House's reassuring words, there was little that Washington could do to get back in the driver's seat.

The American press could hardly resist drawing parallels between oil prices at the turn of the millennium and in the crises of the 1970s. Oil was back in the headlines—and for all the wrong reasons, at least as far as most drivers were concerned. Still, mainstream oil analysts chafed at the thought that Colin Campbell and his pessimistic peers were being vindicated by this turn of events. Two prominent analysts, writing in *Foreign Affairs* in early 2000, expressed little patience for the doomsayers. If anything, they argued, Campbell and his fellow "shortage theorists" had it exactly backward. "The energy problem looming in the early 21st century," they wrote, "is neither skyrocketing prices nor shortages that herald the beginning of the end of the oil age." Rather, the problem this time was precisely the opposite one: a persistent oil glut, which would stretch over the next two decades and could spell disaster for oil-producing countries.

The authors singled out Campbell as an example of the Chicken Littles who were getting it all wrong. A parade of earlier "scarcity theorist[s]" had come and gone since the 1970s, they noted. Now their descendants were wrong-headedly arguing "another energy crisis" was in store. They gave no credence to his conclusion that "by 2003, global production will peak and the oil age will start to end." Such pessimism was only the latest example of a "'sky-is-falling' school of oil forecasting" that had proven "systematically wrong for more than a generation." They extolled the relentless innovation underway in the oil industry, which had made oil cheaper and easier to find than ever before.[24] As they saw it, the world was entering a generational buyer's market for oil. Not only was there more than enough oil to go around, but the abundance would keep prices at all-time lows as far as the eye could see.

Unfortunately, the oil market in the year 2000 did not cooperate with their thesis. OPEC producers remained unusually disciplined in keeping oil off the market in a bid to keep prices within their preferred range of $25 to $28 per barrel. Business inventories of gasoline and diesel were far lower than unusual. Spot WTI prices whiplashed between $24 and $34 per barrel throughout the first half of the year. Prices above $30 per barrel were extraordinary enough in their own right; coupled with such volatility, the market was as turbulent as it had been in a decade, and far more persistently. "Why $30 Oil?" the headline of the IEA's monthly *Oil Market Report* blared. "What's going on here?" Although they saw the market as being well supplied, IEA economists foresaw trouble over the rest of the year, and that worry was feeding back to higher prices today. They pointed to "concerns about the adequacy of gasoline supplies this summer and heating oil next winter, doubts about the durability of Iraqi output" and skepticism that OPEC would stick to its sub-$30 price band for much longer as all contributing to the surprisingly bullish market.[25] At their September meeting in Vienna, OPEC ministers agreed to raise production quotas by 800,000 barrels per day in a bid to get prices to slip back under $28 per barrel.

But chance events and ever-rising demand later that year did not oblige. By August, U.S. inventories of crude oil had fallen to twenty-four-year lows. Harsh rhetoric between Iraq and Kuwait just days after the Vienna summit roiled the market anew, sending NYMEX crude over $37 at the close of the day's trading session on September 20. Gasoline and heating oil prices in the United States were at their highest point in a decade. It was enough to convince President Clinton, heeding then-Vice President Al Gore's campaign-trail call for "aggressive national action" to lower oil prices for U.S. consumers, to tap national stockpiles in an attempt to lower prices. The White House authorized the release of 30 million barrels of crude oil from the Strategic Petroleum Reserve over the course of thirty days. The release took the form of a swap, whereby companies could borrow oil immediately for a promise to replace it later, by which point the market had softened, fingers crossed. Republican presidential nominee George Bush decried the move as "bad public policy," but there was little he could do to stop Clinton's attempt to pacify the market.[26]

The Strategic Petroleum Reserve release reinstated order in the market momentarily, but chaos elsewhere re-injected turmoil. A terrorist attack on the *USS Cole* in the Yemeni port of Aden sent the price of NYMEX crude oil for delivery the following month soaring to $36 per barrel. Although OPEC announced on October 30 that it would raise quotas to try to take the edge off prices, many analysts doubted how achievable the 500,000–barrel–per–day increase really was. After all, most OPEC members were already pumping all-out. OPEC ministers reversed course the following month, however. The move marked an end to any pretense of the cartel trying to uphold the $28 upper limit to prices they had once professed. What was more, Iraq suspended its exports temporarily in early December as an act of defiance over its oil-for-food program administered by the United Nations. Surprisingly, though, the market shrugged at the loss. In a final bout of schizophrenia, oil prices fell 28 percent in the final month of the year, with spot WTI sliding to $26 per barrel as global oil inventories built. Times were strange in the oil market. But they would get much stranger in the decade ahead. In its final monthly report of the year 2000, the IEA ominously reminded readers, "We are not out of the woods yet."[27] In fact, they were just stepping into them.

With the market reeling, the notion that the world might soon bump up against an absolute limit in how much oil it could produce began to creep into mainstream debate. There was no question in any camp that the oil infrastructure in place was close to being maxed out. The flow of oil from OPEC and other major producers, like Russia and the United States, was chronically disappointing. The more controversial issue, however, was how much rates of production could increase in the future.

One of the first prominent groups to wade into the debate was the U.S. Geological Survey (USGS), which released an exhaustive report on the world's oil reserves in early 2000 compiled by a team more than forty geoscientists. The report aimed to gauge the amount of conventional oil outside the United States with the potential to be added to global reserves to 2025. As the USGS saw it, the cumulative production of conventional oil outside the United States was nowhere near reaching its midpoint, as Campbell and peak-production theorists argued. Federal geologists estimated that some 539 billion barrels of oil equivalent (mean estimate) had been produced between the beginnings of the industry and 1995. But this oil was dwarfed by what remained: 859 billion barrels of remaining reserves, two-thirds of which were relatively easy-to-extract conventional reserves, and 649 billion barrels of conventional reserves still undiscovered. Easy-to-access oil was not halfway gone. Emphasizing that its estimates were the product of "geologic studies as opposed to statistical analysis," the government scientists subtly yet pointedly sought to distinguish their approach from their peak-minded brethren, who believed that future production would adhere to their smooth, symmetrically drawn curve.[28]

Drawing on the USGS study, the Department of Energy's hub for oil market analysis, the U.S. Energy Information Administration, weighed in April 2000. At a conference of the American Association of Petroleum Geologists in New Orleans, EIA Administrator Jay Hakes gave a PowerPoint presentation whose title spoke to the million-dollar question facing the oil industry: "The Year of Peak Production: When will worldwide conventional oil production peak?" Hakes started with the obvious: "The prospect that world oil production might actually start to decline in the next ten years is indeed disturbing." He acknowledged that some forecasters on both sides of the Atlantic were putting "this pivotal world event" as arriving as soon as between 2004 and 2010. But did the Department of Energy agree?

Not really, Hakes explained. For one thing, Campbell and Laherrère were likely far too low in their estimation of the world oil resource base, which they put at 1.8 trillion barrels of recoverable oil. The EIA put its mean estimate at just over 3 trillion barrels, and it was all but certain there was at least 2.2

trillion barrels. He astutely noted that published estimates of the world's total amount of oil in the ground had tended to go up over the decades, thanks to more and better data about below-ground deposits. The best estimates of the 1940s had pinned total resources at about 600 billion barrels; sixty years later, U.S. geologists calculated it was as high as eight times that (and counting). In its view, the peak year of production was likely sometime between 2030 and 2075. Whether it came at the near or far end of that interval depended largely on how quickly oil production grew year to year. Hakes acknowledged that "market feedback mechanisms"—that is, the interplay of supply with price and demand—might smooth or flatten any eventual peak. Other factors, like the discovery of more conventional oil sometime in the future or a change in global oil demand, would also matter to production patterns.[29] In short, not only was peak oil not imminent, but the theory itself was far too simplistic to be of much help in predicting the path of a complex, market-driven phenomenon like oil production.

The USGS report flew in the face of peak oil's foundational claim—that the world's conventional oil was halfway gone. It was enough to cause the International Energy Agency to throw out the peak-ish supply forecast it had published two years earlier in its annual World Energy Outlook. The 1998 IEA report had taken a page out of Hubbert's peak oil paradigm by trying to guess at future production rates based on an estimate of ultimate recoverable oil reserves. Drawing on estimates of the world's total oil resource base from a 1993 USGS report, the IEA predicted that the world's oil production would stall sometime between 2010 and 2020.[30] Unconventional sources like rock-like shale oil and ultra-heavy oil from Canada and Venezuela might help ease the pain when peak oil hit, but only at much higher prices. The 2000 USGS numbers helped convince the agency to change its mind. The IEA decided that maybe the peak was not imminent after all. "One need expect no global 'supply crunch,'" analysts wrote in a new version of its flagship Outlook. "The resource base of the world as a whole is not a constraining factor." Investment was a bigger bottleneck than geology. Bringing the oil out of the ground and to the market would be expensive, demanding "significant and sustained capital investment." But OPEC and unconventional oil could help fill any declines in production elsewhere. The Paris-based agency assumed that international oil prices would stay flat at $21 per barrel over the next decade, gradually rising to $28 per barrel through 2020.[31]

The New American Energy Crisis

For most people, these were arcane debates. The end-of-oil idea was not yet the stuff of headlines. Yet oil was on the national radar in a major way, for one simple reason: gas prices were giving drivers fits, and politicians were getting an earful from their constituents. By February 2000, prices at the pump had climbed to their highest in a generation in nominal terms—a whopping $1.38 per gallon. They would stay north of $1.40 per gallon for the next twenty months, a plateau not sustained since the Saudi oil crash of 1986.[32] President Bush's economic advisors fretted that such a rapid spike in prices might be enough to tip the economy into recession, recalling the doldrums of the 1970s, though others felt that progress in energy efficiency since that time provided at least a partial shield.

Some in Washington were getting worried. The nation was facing nothing short of an "energy crisis," President Bush's newly appointed Secretary of Energy, Spencer Abraham, said. "The world has not run out of oil or other resources," he told a national energy summit in March 2001. "Yet, here we are, with the most serious shortage since the days of oil embargoes and gas lines."[33] A recent spate of electricity blackouts in California had made things appear all the more ominous. For Carl Levin, a Michigan Democrat who assumed the helm of the Senate Permanent Committee on Investigations in June, such pronouncements were not enough. "The oil companies need to explain why gas prices have increased so dramatically," Levin told an impatient public.[34] The senator was not the only one concerned. Energy was "America's new number one problem," a May 2001 Pew poll found. Energy problems had "captured the attention of the American public to an extraordinary degree," with more people citing energy as the nation's "top concern" than they had any other single issue since the mid-1990s.[35]

In the summer of 2001, $30-per-barrel oil seemed to many Americans like more than they could bear. But it was only the beginning. Between 2001 and 2003, the market was about to be struck by the disruptive force of geopolitics with a force nearly unimaginable in the relative tranquility among oil producers in the latter half of the 1990s. The first rupture would occur on September 11, 2001, the day of the deadliest terrorist attack in U.S. history. Other than briefly sending oil prices soaring, the attack had little direct effect on the industry. The NYMEX, shuttered amid the devastation, resumed trading on September 17. Oil prices actually declined over the next few months, falling below $20 per barrel until March 2002, as traders weighed the possibility of a recession biting into demand for energy.

But the shock waves from September 11 would affect the oil landscape in other ways. For one thing, it exposed the festering instability and rampant anti-Americanism in the heart of the oil-producing world, the Middle East and North Africa. There had been other terrorist attacks before: on the World Trade Center in 1993, for instance, and later on the embassies in Kenya and Tanzania. None of these were anywhere near as worrying to oil traders, though, as the September 11 attack. Such a vivid and potent act of violence, emanating from the heart of the oil-producing world, was enough to give the oil market serious pause about what other risks might be lying beneath the surface. It was now tragically obvious to the world that the deep domestic unrest in the Middle East had the capacity to wreak havoc on the global energy market. The attack also added an element of unquiet in the most important oil trade relationship of the twentieth century, the one between Riyadh and Washington. No fewer than fifteen of the 9/11 hijackers were Saudi. Later attacks by Al-Qaeda-linked operatives on Saudi targets showed that both countries had common cause in the war on terror, but the event marked the beginning of a more complex chapter in their partnership. For traders, the event was an unmistakable signal that hidden risks in the oil market were lurking. The events of 9/11 would also darken American public opinion about oil. Ending the country's "addiction to oil," in hopes of insulating it from the problems of the Middle East, became a rallying cry. The quest for U.S. "energy independence" became a cultural meme and a strategic priority with an urgency it had not seen since the days of Jimmy Carter.[36]

Yet much nearer the home front other disruptions to the old order of oil was also underway. Trouble was brewing in Venezuela, South America's largest oil producer. Hugo Chavez, a charismatic career military officer who had been elected president in 1998, had begun to clamp down on his opposition. No sooner was he reelected in 2000 than he began to take steps to solidify his control over the oil-wealthy nation, including passing a series of laws that magnified his control over the state without the consent of the National Assembly. But it was the country's national oil company, Petróleos de Venezuela, known as PDVSA, which was the real prize. In April 2002, Chavez summarily fired seven members of the board of PDVSA. For his opponents, this was a step too far. Four days later, more than a million people took to the streets in Caracas, ultimately leading to a failed coup attempt. But his opponents were undeterred. That December, they launched a general strike, which extended to PDVSA. Venezuelan oil production quickly collapsed from around 3 million barrels per day to around 600,000 barrels per day. By early 2003, much of the oil was flowing again. But Chavez was determined that it would never happen again. Eighteen-thousand PDSVA workers were let go, replaced by Chavistas who lacked the skills or experience of their predecessors. Caracas began dipping into the company's coffers to fund pet projects with abandon. Once one of the world's exemplary national oil companies, PDVSA soon became unable to maintain its former production levels, let alone meet its growth targets. The country's production would slowly decline over the next few years, adding strain to a world oil market that badly needed more oil.

"Nature Is Not Making Any More Oil"

It was in these years—the shadow of 9/11 and collapse in Venezuela—that peak oil started to become more than just an idea—it became a movement. For one thing, it was picking up adherents of greater prominence. One was Kenneth Deffeyes, a Princeton professor emeritus of geology, who had been a protégé of Hubbert at Shell. His 2001 *Hubbert's Peak: The Impending World Oil Shortage*, published by the well-regarded Princeton University Press, argued that world oil production would peak "in this decade—and there isn't anything we can do to stop it." The geology professor admitted that Hubbert's legendary 1956 prediction that U.S. oil output would soon peak was "just barely within the envelope of acceptable scientific methods. It was as much an inspired guess as hard-core science."[37] Yet it was good enough for Deffeyes to broadcast it anew. Deffeyes's impressive academic credentials lent an air of respectability to the peak oil movement. "A heavyweight has now joined this gloomy chorus" of oil skeptics, the *Economist* wrote.[38] Other heavyweights followed. Matthew Simmons, the founder of the largest oil-focused investment bank, Simmons and Company International, became an indefatigable defender of the faith, perhaps its most notable champion from the ranks of industry. Cal Tech physicist David Goodstein, crunching the numbers for himself, also became a believer. By 2007, he argued, world production could hit its peak. "Nature is not making any more oil," the professor reminded the public.[39]

Research institutes and professional conferences focused on peak oil also began to pop up. The Oil Depletion Analysis Centre was launched in the United Kingdom in 2001, backed by donations from the Astor family. Its mission was to "raise international public awareness and promote better understanding of the world's oil-depletion problem."[40] That November, it brought together what one journalist described as a "colorful group" of several dozen people at Imperial College to hear Colin Campbell explain his provocative decline curves. Campbell did not disappoint, decrying the "amazing display of ignorance, deliberate ignorance, denial and obfuscation" that government officials, academic experts, and oil industry bigwigs had shown in their denials of peak oil. Other peak oil groups also came together. The Association for the Study of Peak Oil, or ASPO, grew out of a talk that Campbell had given at a small university town in Germany in December 2000. Campbell found a highly receptive audience among European scientists and researchers, who began to organize informally to discuss the issue of worldwide fossil fuel depletion. Jean Laherrère eventually convinced the group to add "natural gas" to the name, though it stayed ASPO for short. The organization's aim was to "raise awareness of the serious consequences of humankind" of energy depletion. ASPO stepped onto the public stage by staging the International Workshop on Depletion at Uppsala University in Sweden in May 2002.[41] The conference made headlines. "Crude Oil Supplies Could Peak by 2010," an Associated Press write-up of the event blared.

The Revenge of the Old Economy

Some far-sighted market analysts, though not members of the peak oil crew, had also begun to sense that oil prices were indeed about to rise, perhaps even explode. The "revenge of the old economy," Goldman Sachs commodity analysts explained in an April 2002 note, was about to strike. "Years of underinvestment in commodity-industry capacity means that the global economy may run out of capacity in core commodity industries, including energy," the bank told investors. The ultra-low oil prices of the late 1990s, driven by the Asian financial crisis, had halted the flow of new investment into the energy sector. This lack of investment meant that the funds necessary to maintain the productivity of oil and gas fields, put new drilling rigs in operation, and expand global refining capacity (which had retreated to 1982 levels) simply were not there. This missing investment would be "keenly felt" once economic growth regained momentum. Higher prices were in the offing, Goldman told investors, "likely to rise sooner, and to a higher level, than is typical."[42]

Another prominent voice in the commodity markets, Andy Hall, a veteran trader at Phibro, an energy trading house owned at the time by Citigroup, agreed. By late 2002, Hall had become convinced that the market was grossly underestimating how expensive oil would be in a few years' time. The sharp divergence in opinion about future supplies among the "irreconcilable optimists, wishful thinkers, pessimists and alarmists" had nudged him to "get thinking" about where the market was headed. "I've dug into the facts," Hall told a group of Citi executives in New York. "And saw that supply will not meet demand." He outlined his case for a massive "three- to four-year bet" that prices would rise—perhaps even testing $100 per barrel. "Everyone else is behind the curve," the trader declared.[43] Neither the analysts at Goldman Sachs nor Hall were making their case for a rise in prices on the claim that global oil resources were half gone, as peak oilers were, nor did they see the imminent price rise as never-ending. But they did share at least one thing in common: a conviction that prices were indeed going higher, and soon.

Their timing was spot on. Spot Brent was trading at less than $20 per barrel at the end of 2002. A year later, it closed at just over $30 per barrel. The market was struggled to cope with what the IEA called "unprecedented risk." The list of global hot spots was long: the unfolding crisis in Venezuela, resurgent

petro-nationalism in Russia, kindling Israeli–Palestinian tensions, and the
threat of confrontation between India and Pakistan. Cold winter weather had
juiced the market in the fourth quarter of the year. Industry stockpiles of oil
were thin on the ground in the United States. Even more troublingly, the
amount of slack in the system was looking painfully low. By the outset of 2003,
spare production capacity had slipped below 2.5 million barrels per day—less
than one-third the buffer that had existed a year earlier, not counting constrained supplies from Venezuela and Iraq. Surging to make up for lost supplies from Caracas, Saudi Arabia was a mere 1.5 million barrels a day from
pumping at what generous estimates believed were its max-out levels. But
these were not the only factors pushing up prices over the course of 2002.
There was one additional variable: "concerns about Iraq," as the IEA would put
it.[44] As 2003 unfolded, these concerns would turn out to be well placed.

"Running on Empty"

On March 20, 2003, U.S. forces invaded Iraq. The invasion was the culmina-
tion of deliberations within Washington since the 9/11 attacks about how to
respond to the prospect of a Middle East whose violent unruliness had become
a threat to American lives and global stability. The war sprung from several
factors: the threat of terrorism in the aftermath of 9/11, the risk of weapons of
mass destruction, and Saddam Hussein's staunch opposition to the West
among them. It was more than clear to many in the oil market in the months
preceding the invasion that the United States and its allies, dubbed the "coali-
tion of the willing," might be preparing to go to war. U.S. officials knew that
upsetting the country's oil economy (and thereby sending prices higher) was a
major risk, but State Department policy planners prior to the war had "extremely optimistic expectations about how quickly production and exports
could be restored," according to one account. One senior Department of Defense official had estimated that a postwar Iraq could soon pump 6 million
barrels a day, in contrast to the 2.5 million barrels a day at the time of the
invasion.[45]

Such optimism about Iraqi oil production turned out to be unfounded, at
least over the short term. Just a week before the invasion, the IEA issued a
warning about the fragility of the market via its monthly *Oil Market Outlook*.
The oil market was "running on empty," and any additional turmoil among oil
suppliers would likely end badly:

> Industry oil stocks are tight, and trending around minimum operat-
> ing levels in key markets. Surplus production capacity is low due to

capacity losses in Venezuela and an offsetting surge in OPEC supply. And current oil prices are high, fuelled by low stocks, weather-related demand and anxiety about military intervention in Iraq. A further supply disruption would tax a system operating at close to capacity.[46]

Yet "supply disruption" was exactly what was about to occur, as sanguine predictions about an Iraqi production resurgence turned out to be woefully overoptimistic. In the immediate aftermath of Baghdad's fall, the country produced next to nothing, a measly 53,000 barrels per day. It would not be until January that production would get back to 2 million barrels per day. Beset by security challenges—bombings and sabotage were rife—not to mention antiquated, inadequate infrastructure and technology, Iraq's oil industry would struggle in the seven years following the invasion just to hit the 2 million barrel mark. Only in the best of times could the country maintain production at two-thirds its IEA-deemed capacity, let alone actually grow that capacity. Not until July 2008 would Iraq produce as much oil, on a monthly basis, as it did just before the United States and its allies put boots on the ground.[47] Having one of the world's most prolific reserve holders struggling to keep its head above water was not a welcome development for a market already on edge.

Booming China and a "Commodity Supercycle"

Underinvestment in new production in the late 1990s and trouble in OPEC in the early 2000s were setting the stage for a boom. But it was demand that lit the fuse. Beginning around 2002, the world economy moved into a higher gear. It had grown just a hair over 3 percent over the 1990s; between 2002 and 2007, growth was just shy of 4.5 percent, nearly 50 percent stronger.[48] And unlike prior periods of rapid economic growth, this one was truly global in nature, hardly limited to the advanced economies of the United States, Japan, and Western Europe. Faster growth meant more appetite for oil. The numbers are stunning. Over the entire course of the 1990s, oil consumption outside the OECD grew by less than 3 million barrels per day. It would eclipse that figure between 2003 and 2005 alone. On a global scale, demand grew nearly three times as much between 2003 and 2006 as it had between 1999 and 2002.[49] What was more, the map of global oil consumption was shifting toward new epicenters—in the populous developing economies of Asia (China and India, in particular), the Middle East, Eastern Europe, and Latin America. Generous government fuel subsidies in many of these areas meant that consumers were insulated from high oil prices elsewhere. Demand for oil among a new, rapidly urbanizing middle class, hungry for the comforts of Western life, was soaring.

The 800-pound gorilla among these new oil buyers was unquestionably China. The Middle Kingdom was on a tear. In the five years leading up to 2003, its economy had logged 8.3 percent average growth; in the five years that followed, it would catapult to nearly 12 percent.[50] Developing the economy, rather than staying true to the old communist doctrines, had become the country's rallying cry. Global trade was accelerating, and low-cost Chinese factories were a hub for building, assembling, and exporting a vast array of finished goods. Chinese workers, flocking to its coastal cities, were getting richer. GDP per capita more than doubled between 2000 and 2006. They were also taking to the streets. Car sales would also double between 2002 and 2006, then nearly double again by 2009, the very year that sales in the China overtook the United States—a tilting of the industry almost unthinkable a decade earlier.[51]

Faster growth, bigger cities, richer consumers, and more cars—taken together, they all meant more oil. Not only were the country's economy and cities growing rapidly, but they were growing in a commodity-intensive way, requiring massive amounts of energy and other raw materials to propel their export-driven progress. Coal, the country's dominant energy source, simply could not continue to provide for the country's needs. Demand for it was growing, too, of course—the increase in Chinese coal consumption alone between 2000 and 2010 would account for a mind-bending 40 percent of the entire world's growth in energy used. But oil, more and more, was figuring into the mix, both as a means of generating electricity, whether for diesel-powered generators or power plants, and to fuel cars and trucks. The Asian giant accounted for one-third of all additional oil consumption between 2001 and 2007. Its 1-million-barrel per day growth in a single year, 2004, stood shoulder to shoulder with all of its gains over the entire second half of the 1990s.[52] A "commodity supercycle," as one Wall Street analyst would later term it, was underway, propelled by the engine of Chinese demand.[53] This was no ordinary boom in natural resource markets—it was the beginning of a massive bull run.

Had oil companies been able to foresee such remarkable growth in the years prior, they might have been able to devote the resources necessary to expand their production capacity before the demand boom. But they were caught largely unawares. As late as the summer of 2004, some executives at the world's biggest oil companies were gun-shy about sinking capital into projects that could not balance at their "long-term price guidelines"—in the low $20s. Fearing another 1998-style price collapse, oil companies chaffed at the thought of overextending themselves in new exploration and production initiatives, only to find themselves stuck with uneconomic projects if prices—or "when prices," in their view—reverted to their long-term mean.[54] By the time it was obvious that China and its peers in the developing world were undergoing a massive growth spurt, and that the $22–$28 price band that OPEC had once stood behind was now a thing of the past, all oil suppliers could do was play catch up.

Growth in oil production outside of OPEC's borders struggled to help satiate the climbing demand. The numbers that countries like the United States, Canada, China, Mexico, and their brethren were putting up did not bode well. Despite prices that provided ample incentive to pump oil, production outside OPEC stagnated at around 46 million barrels per day between 2003 and 2007. The former Soviet Union was a bright spot, putting up nearly 750,000 barrels per day of growth in the first half of the decade. Without that boost, non-OPEC production was actually slowly declining. The 35.9 million barrels per day in 2002 slipped lower in 2003, then in 2004, and again in 2005, in a pattern that would continue for the next five years. Although OPEC could—and did—raise its production to offset the demand for more oil, it did so largely through eating away its own excess capacity, which mid-decade had worn to a fraction of what it had been at the start of the decade. Building new capacity was slow, and maintaining production capacity that went unused was costly. Between 2000 and 2006, OPEC's effective production capacity barely budged, never straying far from 32 million barrels a day (except for when it temporarily plummeted in 2003 due to troubles in Venezuela and Iraq).[55] Producers outside OPEC were pumping nearly all out, and now OPEC appeared to be headed in that same direction. For a world that badly needed more oil, it was a pattern that could only mean one thing: higher prices.

A "New Era" of "Irreversible Decline"

For disciples of peak oil, however, it meant even more than that. It meant vindication. Year after year, and despite rising prices, the production data disappointed, just as they had foreseen. It looked as if the world really had begun to bump up against the geologically determined limits of how much oil could be called forth. The bright future for world oil supplies that their opponents had foreseen was starting to look more and more dubious, at least to them. It was all just as they had said it would be, and at the very time that many of them had said it would happen. Hubbert himself, after all, had foreseen a global production peak around the year 2000. The new millennium was now several years underway, and the legendary geologist was looking more prophetic than ever.

As 2003 gave way to 2004, there was little question that the market was tightening. Inventories of crude and refined products dwindled over much of 2003, trending below their five-year averages. OPEC surprised analysts by deciding at their September meeting to cut production, fearing that the market could be oversupplied the following year, and acting out of an abundance of caution. In truth, there was plenty of reason to doubt that prices were set to fall, with civil unrest in Nigeria making waves, Caracas continuing to simmer, and "placid comments," as the IEA put it, from producers outside of OPEC with regarding to their own production plans for 2004.[56] Yet the cartel was acting with the memory of 1998 firmly in mind, on guard against opening the taps too liberally when resurgent Iraqi exports and a looming seasonal demand slowdown could drag down their revenues. As it turned out, though, tepid autumn demand was the last thing they needed to worry about. Oil consumption, especially in China, was roaring ahead of IEA estimates. In the spring, OPEC went ahead with production cuts agreed upon in the fall. Two deadly terrorist attacks in the heart of Saudi Arabia in April and May added to market jitters. If OPEC's aim had been to put a floor under prices, they were more than successful. By summertime NYMEX prices for crude to be delivered the following month reached an all-time nominal high of $42.33 per barrel. Oil was soaring.

Yet peak oil's skeptics were not about to concede that the "big rollover" had actually arrived. One of the idea's most prolific antagonists was Michael Lynch, an outspoken MIT-trained economist and energy gadfly. He had written contemptuously of the peak oil idea ever since Campbell and Laherrère's *Scientific American* article had generated buzz in 1998. "The profusion of articles" claiming the "end of the oil era" is "unfortunate," Lynch had written at the time, "since the casual reader (and policy-maker) might conclude that the large number of articles have an equally large amount of research behind them." But predictions that oil production would soon enter terminal decline "due to inexorable geological forces" were pure silliness, owing to faulty resource estimates and bad economics. "Short-term prices will certainly fluctuate, and we will surely have more oil crises," Lynch acknowledged. Yet with any luck, the "certain failure of the current Cassandras will be forgotten within a few years," and not "receive the attention that the current (and previous) ones did" in the years to come.[57]

Lynch kept up his ridicule even as prices soared over the course of 2004. "Oil production in various countries has flattened or fallen at certain times for reasons having nothing to do with how much they could produce," he told peakists. "Hubbert modelers have not discovered new, earthshaking results but rather joined the large crowd of those who have found that large bodies of data often yield particular shapes, from which they attempt to divine physical laws."[58] For Lynch, peak oil was nothing more than old wine—and bad wine, at that—in new bottles.

Rising prices did not mean that peak oil had actually struck, Lynch and like-minded skeptics continued to insist. Amy Jaffe, whose 2000 essay for *Foreign*

Affairs had envisioned oil prices at rock-bottom levels far out into the future, ate crow with a postscript in the same magazine in May 2004. "Forecasts that long-term oil prices would recede again this decade have turned out to be well off the mark," she conceded, "but not because those who had predicted scarcity had it correctly." Rather, the problem was "fundamentally political." There were vast amounts of oil under Russia, Saudi Arabia, Venezuela, and West Africa, certainly enough to create a long-term bear market. But war and rumors of war had interceded. The current run-up in prices "will probably be solved, as it has repeatedly over recent history," she wrote, striking a far more tentative tone. But the price collapse may not come about without "a good sized recession that deeply curbs demand."[59] In other words, peak oil was not real, but a painful price rise was looking increasingly possible.

Others were not so sure. It was all but taken for granted in mainstream energy circles that while countries like the United States and Britain had indeed passed their prime, vast Middle Eastern suppliers like Saudi Arabia still contained almost unthinkably enormous quantities of oil. Now that notion was starting to come under fire. A provocative article in the *New York Times*, run in February 2004, made the case that Saudi oil resources—always shrouded in mystery among oil analysts and treated as a state secret by Riyadh—were not nearly as abundant as people believed:

> When visitors tour the headquarters of Saudi Arabia's oil empire—a sleek glass building rising from the desert in Dhahran near the Persian Gulf—they are reminded of its mission in a film projected on a giant screen. "We supply what the world demands every day," it declares.
>
> For decades, that has largely been true. Ever since its rich reserves were discovered more than a half-century ago, Saudi Arabia has pumped the oil needed to keep pace with rising needs, becoming the mainstay of the global energy markets.
>
> But the country's oil fields are now in decline, prompting industry and government officials to raise serious questions about whether the kingdom will be able to satisfy the world's thirst for oil in coming years.

The article cited various executives, some named and others not, claiming to have inside information on the kingdom and its massive national oil company, Saudi Aramco. Edward O. Price, Jr., a former top Saudi Aramco executive, admonished that "the world should not expect more" than 12 million barrels a day of production, suggesting that the country's ability to supply the market would soon hit a wall. An unnamed Saudi official warned that even trying to reach those production volumes would "wreak havoc" within a decade. "In the past, the world has counted on Saudi Arabia," a "senior Saudi oil executive" was quoted as saying. "Now I don't see how long it can be maintained."[60]

He was not alone. Matthew Simmons, by this time a well-established peak oil authority, was also transfixed by the notion of a looming Saudi oil shortage. In the summer of 2005, Simmons published *Twilight in the Desert: The Coming Saudi Oil Shock and the World Economy*, which made an elaborate case for why the kingdom's oil production was fast approaching an "irreversible decline."[61] Saudi fields were being flogged to death, he argued, and unsustainably high rates of production meant that the kingdom would be lucky if it could even sustain current levels. Peak oil would soon strike in the heart of the oil world. The book's bold claims made it a bestseller, though it was impossible to verify at the time how much merit there was to Simmons's contentions without the Saudis releasing their geological data to the world. Riyadh's penchant for privacy (the kingdom considers the details of its oil resources a state secret) meant that it was hard to confirm or deny the book's claims about the technical problems its oilfields were purportedly confronting. Either way, the book provided plenty of grist for the idea that peak oil was about to infect one of the few producers most everyone thought was immune—and with potentially disastrous results for world oil supplies. "If Mr. Simmons is right about Saudi Arabian oil production . . . we can kiss the era of abundant petroleum goodbye forever," wrote one reviewer.[62]

Oil prices continued to race higher, soon breaking into territory not seen in decades. A massive hurricane, dubbed Ivan, rocked the Gulf of Mexico in September, the biggest disruption of the region's oil infrastructure in years. More than 3,000 workers were evacuated from offshore platforms as oil companies hunkered down. Nearly two-thirds of the region's oil production was temporarily lost. By February 2005, it would become clear that the hurricane had removed over 7 percent of annual production from the Gulf. There was trouble elsewhere, too. November saw saboteurs blow up three major pipelines in northern Iraq, taking crude exports via Turkey offline for three days. In Nigeria, around 300 unarmed villagers took control of three oil flow stations in the Southern Niger Delta, the second episode of unrest in the area in two weeks' time. Protesting a lack of jobs, the villagers were successful in shutting 100,000 barrels a day of production in for a week. Unsurprisingly, global oil markets reacted fitfully to all the bad news. Futures prices for WTI crude oil trading on the NYMEX surged to an all-time high of $55.17 in the fall. "This is a new era," Matt Simmons told an oil industry crowd at a conference the following April.[63] If the action on the NYMEX was any indication, it certainly seemed to be.

A controversial piece of research commissioned by the U.S. Department of Energy added fuel to the fire in February 2005. Federal energy officials hired Science Applications International Corporation, or SAIC, a research firm, to do a deep-dive into the question of peak oil. Robert Hirsch, a former Exxon analyst and director at the Atomic Energy Commission, quarterbacked the effort with the help of two other investigators, which culminated in an exhaustive 92-page report, *Peaking of World Oil Production: Impacts, Mitigation, & Risk Management*, commonly known as the Hirsch Report. It did not mince words. "The peaking of world oil production presents the U.S. and the world with an unprecedented risk management problem," the authors intoned. It soberly presented a list of "competent forecasters" who projecting "peaking within a decade," while acknowledging that it was a complex issue. Still, "peaking will happen," even if the "timing is uncertain." Hirsch and his co-authors tried to impress upon readers' minds the severity of the imminent crisis. "The world has never faced a problem like this," they wrote. "Oil peaking will be abrupt and revolutionary." Governments needed to act, and fast, or else things could go badly wrong. Unless governments intervene, "the economic and social implications of oil peaking" will be "chaotic." It was not simply a matter of oil becoming problematically expensive—an absolute shortage could well be on the way. Cars, airplanes, trucks, ships—all moving things that ran on oil and could not switch to other energy inputs—could come to a halt . . . for good.[64] From a report funded by Washington, and apparently written at the behest of the country's top energy thinkers, it was an alarming message.

Venerable Wall Street voices continued to predict that oil prices were going higher—much higher, in fact. Goldman Sachs, the same firm whose "revenge of the old economy" hypothesis had made waves a few years earlier, decided to double down on its buy recommendation of a few years earlier. In March 2005, Arjun Murti, an analyst covering integrated oil and gas companies, put out a note to clients with a message that immediately went viral: Get ready for triple-digit oil prices. "We believe oil markets may have entered the early stages of what we have referred to as a 'super spike' period—a multi-year trading band of oil prices," the Goldman team wrote.[65] Unlike Simmons, the team at Goldman did not weigh in on peak oil one way or the other. But they did judge that the crisis-level oil prices of the 1970s might be in the offing. Other analysts at the world's major banks thought Murti had gotten out over his skis. "Of course we could go to $105, if Saudi Arabia's reserves were destroyed," a Wachovia analyst told the *Wall Street Journal*. "But the probability of that is so low, it's not worth publishing on the front page of a report."[66] Others implied bad faith. One market watcher told reporters that Goldman was just talking its book, trying to "talk up the market on nothing more than hot air" to juice their own trading profits.[67] Few disputed that the oil market was hot, and getting hotter by the month, but few analysts believed that it would get $105-dollar-tight any time soon, unless a disaster struck.

A spate of bad news over the course of 2005 and into 2006 kept prices charging upward, only occasional pausing to breathe. The demand shock of the prior year emanating from Beijing left producers and refiners with little room for error. Outages were rampant: another Iraqi crude pipeline, which ran to the Mediterranean; a harrowing explosion at BP's Texas City refinery, claiming fifteen lives and injuring more than seventy people (not to mention setting the gasoline market on tenterhooks); a strike in France briefly shutting down 90 percent of Total's refining capacity nationwide. Things were ugly enough to push WTI prices above $60 per barrel by July. But it would get worse. A deadly wave of storms pounded the Gulf of Mexico in July and August, wreaking havoc on oil production. Hurricane Dennis, a bruising category four storm, disrupted BP's Thunder Horse project. A week later, Hurricane Emily struck in Mexican waters, shutting in the lion's share of the country's local offshore production and causing Pemex to evacuate their workers in the Gulf of Campiche, where the country gets over 80 percent of its oil. But the worst storm would strike in late August. Hurricane Katrina, a category five howler, would wreak havoc on Gulf coastline. All but a drop of oil produced in federal waters had to be shut in as Katrina raged; vital infrastructure was also locked down, and some 2.2 million barrels a day of refining capacity was taken offline. Beyond its human toll, the hurricane was also the costliest disaster in U.S. history. By late September, WTI oil futures had swelled to $70 per barrel.

"Halfway through the Hydrocarbon Era"

What had started as a trickle of news coverage about a world running out of oil turned into a flood. Some thought the idea had merit; others thought it was knuckleheaded. But everyone in energy was talking about it. Even prior to Katrina, a Citigroup analyst told Bloomberg News that there was "no question" oil prices would hit $60. "There is a lot of fear and hype about the possibility of us running out of oil, and it has stuck."[68] An exhaustive CERA report batted back the specter of peak oil, arguing that a comfortable supply cushion of 6 to 7.5 million barrels a day could bring prices much lower as early as 2008. "The main risks to our supply exhaustion situation are above ground," CERA chairman Daniel Yergin argued. NBC News described the report as a "counterweight" to "the predictions of some energy experts"—later calling out Matt Simmons and *Twilight in the Desert* specifically—"who in recent years have been publishing books filled with charts and graphs that aim to prove that world oil production is about to peak, if it hasn't already."[69] More and more newspapers tried to help readers make sense of the noise. "On oil supply, opinions aren't scarce," ran one *New York Times* headline in the wake of Katrina. Although the *Times* acknowledged that high prices had been able to call forth ever-greater volumes of oil in the past, it was far from certain that history would repeat itself this time. Yes, more oil had always come, the article closed. "And it will again. Until it doesn't."[70]

Adding to the noise was the fact that even oil industry titans appeared to have sharply diverging views about what lay ahead. Some executives appeared unfazed. Shell Oil Company chairman Lord Browne was adamant that prices were "unsustainably high." A sharp correction may be in store, he warned. His colleague John Hofmeister, head of Shell Oil Company in the United States, echoed his view, describing this "high price cycle" as "artificially inflated."[71] One legendary American oilman, Boone Pickens, could not have disagreed more. Pickens was adamant in declaring that the industry really was running up against the limits of what it could do. "Global oil production is 84 million barrels (a day). I don't believe you can get it any more than 84 million barrels. I don't care what Abdullah, Putin or anybody else says about oil reserves or production." The world's biggest fields were in decline. "It's a tread mill that you just can't keep up with," he explained. Once you "start adding the reserves in these countries, you're not even replacing what you're taking out."[72] Pickens shared Simmons's bottom line: The world was "halfway through the hydrocarbon era"; oil was more than half spent.[73]

In some corners of Washington, fear over a looming disruption to oil supply lines or a catastrophic interruption to OPEC production was palpable. In the summer of 2005, a cadre of former top U.S. officials holed themselves up in a Washington, DC, hotel ballroom to take part in an apocalyptic war game, dubbed "Oil Shockwave." The exercise, put on by two nonprofit groups advocating that the country get off foreign oil, was meant to help policy planners think through a potential response should prices go off the rails. They played out a frightful scenario: ethnic unrest in Nigeria, which led the country's oil exports to collapse; then al Qaeda-backed attacks that destroyed vital energy facilities in Valdez, Alaska, and Saudi Arabia. As the worst happened, oil prices shot higher uncontrollably. The make-believe White House could only stand by and watch, as the U.S. Strategic Petroleum Reserve did little to stop the bleeding. Soon gas prices were at $5.32 a gallon and the world economy had tipped into a deep recession. It was a warning to Washington. Former CIA director Robert Gates, who had played the role of national security advisor, told the *Washington Post* that the "scenarios portrayed were absolutely not alarmist; they're realistic." The country desperately needed to develop cars that run on "prairie grasses, animal waste and other products" to get the United States off foreign oil and protect it from a world oil order that was increasingly fragile.[74]

Elected officials were starting to get an earful from their constituents about rising gas prices. Six months later, on the other side of the city, a House of Representatives energy subcommittee had called in the leading lights of the peak oil movement to help them understand what was going on in the gasoline market. Kjell Aleklett, the president of the ASPO, and Robert Hirsch both warned policymakers that peak oil was here. "We have all been enjoying the greatest party the world has ever seen: the great oil party."[75] But the party was over, they told Congress. At least some legislators were listening. Two members of the House of Representatives, Roscoe Bartlett of Maryland and New Mexico's Thomas Udall, launched the House Peak Oil Caucus to get the word out among their colleagues.[76] Even oil-patch Republicans, considered the most ardent defenders of the fossil fuel industry, were becoming concerned about the country's dependence on expensive foreign oil, which they worried was enabling rogue regimes and terrorism. President George W. Bush put it bluntly in his January 2006 State of the Union Address. "America is addicted to oil," the Texan declared.[77] The United States desperately needed to find alternative sources of energy.

The oil market ebbed and flowed in a one long arc over 2006, with the usual cast of characters weighing in on what it all meant for the future of supplies. Spot WTI prices were trading comfortably around $60 per barrel at the start of the year. By the summer they had climbed north of $75. Over the next six months, however, they fell back to where they had started the year. But the war of words over peak oil continued, and the tide seemed to be slowly shifted toward greater pessimism. Boone Pickens did not let up on those who thought that high prices would work their magic by bringing more oil out of the ground. "Don't count on technology bailing you out," he admonished. "It's too big for technology. You can just do so much with oil in the ground."[78] Even the head of a prominent industry consulting firm who had predicted a new era of cheap oil just half a decade earlier was losing hope. "Worldwide production will peak," he predicted, which would be the "economic equivalent of a Category 5 hurricane." It was a supply crisis that would spawn "energy rivalries, if not energy wars" and usher in an "age of energy insecurity."[79]

Oil executives, for the most part, continued to scoff at the idea of an oil peak, as did some prominent research firms. "The era of easy oil is not over, because there has never been easy oil," Rex Tillerson, the CEO of Exxon Mobil, used to remind listeners.[80] Nor was CERA, long a voice of optimism, convinced that the end of oil was nigh. In a detailed November 2006 report, the forecasting firm saw global oil production growing for at least the next twenty-five years. It pegged the global resource base of recoverable oil at 3.7 trillion barrels and likely to grow, three times what many peak oil advocates assumed. New drilling techniques and technologies would enable companies to tap previously unreachable resources. Production could rise by more than 50 percent to 130 million barrels a day by 2030, after which point it would "undulate" higher and lower for the next few decades at least.[81] "The peak-oil theory causes confusion," CERA analysts wrote. Hubbert's ideas were "misleading" and based on shoddy, overly bleak estimates of oil in the ground. There was no "imminent peak," nor would supplies "run out dramatically soon thereafter."[82]

"High Prices, High Uncertainty, Low Stocks"

However credible their circumspection, undoing the conviction among peak oil advocates that the end of oil really was near was no mean feat. In bookstores and on the Internet, worry over an irreversible decline in oil production was reaching a crescendo. There were enough treatises on peak oil to populate a small library. Some saw peak oil as leading to a catastrophic decline. Titles like *The Long Emergency: Surviving the End of Oil, Climate Change, and Other Converging Catastrophes of the Twenty-First Century* (2006) fit into that category. Others, like *Powerdown: Options and Actions for a Post-Carbon World* (2004) and *The Post-Petroleum Survival Guide and Cookbook: Recipes for Changing Times* (2006) implied that peak oil might actually lead to a cleaner planet and healthier communities. Still other tomes were out-and-out survival guides. If the end of energy was near, best to be prepared. And books like *Peak Oil Prep: Prepare for Peak Oil, Climate Change and Economic Collapse* (2006) and *Plan C: Community Survival Strategies for Peak Oil and Climate Change* (which appeared in 2008) would tell you how. The Internet was also awash with virtual meeting grounds and watering holes for the peak minded. Concerned citizens and energy experts alike could discuss the future of oil at sites like oildecline.com, postpeakliving.com, oilcrisis.com, and drydipstick.com. Scarcity was in vogue, and with oil prices at once–unthinkable levels, the question of energy sustainability was taking up more and more ink, digital and otherwise.

What began as a tightening oil market in 2007 had turned into an all-out bidding war by early 2008. Spot WTI had sold for $60.77 per barrel on the first trading day of 2007. One year on, as 2008 dawned, the price would be just pennies short of $100. Commercial oil inventories measured in terms of daily demand had slid sharply beginning in July 2007, a dangerous sign of demand getting ahead of supply. The usual factors—impressive emerging market demand, refinery outages, bad weather, disappointing production in the developed world, and instability in Nigeria and key Middle Eastern countries—all played a role. So too did a mismatch among the quality of the oil that surplus producers like Saudi Arabia could put on the market, the types of oil that refineries needed based on their operating configurations, and new environmental regulations in the United States and elsewhere that called for lower sulfur fuels. Slowly, over the course of 2007, the gap between global production capacity and demand shrank, setting off a painful round of rising prices.

By autumn, the only force that seemed to be holding the reigns of the market was uncertainty. "High Prices, High Uncertainty, Low Stocks," screamed the headline of the IEA's October market update. The market was facing as complex a set of dynamics, both fundamental and speculative, as most anyone could remember. High prices appeared to be the "rational result of current market conditions and expected future fundamentals," as best the agency could tell. But the fog of war had set in. It was clear that U.S. demand had recently begun to contract—the early stages of the credit crisis and extraordinarily high gasoline prices were doing their work—but what it meant for prices in 2008 was simply impossible to discern. "The crystal ball will clearly remain fogged for some time," they confessed.[83] The only certainty was uncertainty, in other words.

NYMEX WTI prices closed above $100 per barrel in February 2008 for the first time ever. Come March, prices were just shy of $110. Then they jumped to nearly $120 by April. Yet doubts persisted about whether even triple-digit prices would be enough to tame runaway demand and spur additional supplies—what Economics 101 would predict. Oil was still "preposterously cheap," according to Matt Simmons, who figured it could climb as high as $378 per barrel.[84] Goldman Sachs, whose energy analysts had predicted in 2005 that oil would go to $105, also saw higher prices ahead. In March the bank issued a report arguing that oil could hit $175 in the next few years, envisioning "explosive rallies" in prices due to supply constraints.[85] "We don't subscribe to the peak oil view," one of the bank's top oil analysts told *Barron's*, nor did they think that the "world has run out of oil." But it was getting much more expensive to extract.[86] Many experts— from executives at state-owned oil companies to Wall Street oil analysts— saw prices rising to between $200 and $250 in the near future. Driving their ultra-bullish prognostications was a simple observation: record-high prices did not appear to be enough to make global production rise to meet demand growth. Non-OPEC supplies are "seemingly dead in the water," lamented strategists at Barclays Capital. The laws of economic gravity appeared to be hopelessly missing from the oil market of the spring of 2008. The chief economist of the IEA, Fatih Birol, summed up the sentiment. "According to normal economic theory, and the history of oil, rising prices have two major effects," he instructed. "They reduce demand and they induce oil supplies. Not this time."[87]

Oil Dot-Com

Others on Wall Street saw a speculative bubble in the making. Chief among them was Edward Morse, head of commodity research at Lehman Brothers. In a May 2008 note, *Oil Dot-com*, Morse did not mince words: "We are seeing the classic ingredients of an asset bubble," he wrote. Investors were chasing past price gains, erroneously figuring that triple-digit prices were not enough to spur new supplies and destroy demand. Yes, prices may continue to rise for the next few months, "but when peak prices hit, we believe they are also likely to fall precipitously." Oil is a cyclical business, he warned, and "turning points can be sudden, unexpected, and severe."[88] Others agreed. "I think the market is totally insane," Fadel Gheit of Oppenheimer investments told CNN.[89] Robert Mabro, an oil expert at the Oxford Institute for Energy Studies, agreed. "We're in a bubble right now," he opined. "Prices are rising because everyone expects them to do so. We've seen the same thing in the real estate market."[90] And what comes up, he reasoned, must come down.[91]

While analysts debated what lay ahead, prices continued to rise, and the public continued to fret. On the NYMEX, the price of WTI crude raced past $130 per barrel in May, then surpassed $140 before June was over. On July 11, 2008, oil prices reached their historic peak: $147.27. American consumers were paying more than $4 per gallon for gasoline on average nationwide, eclipsing the high watermark of the 1970s oil crises, after adjusting for inflation.

Anxiety over oil was everywhere. Around the world, not only did it look to consumers like oil might run out—it appeared as if it had. A whopping 70 percent of people worldwide believed that the world's oil supplies were running dry, a poll conducted by the University of Maryland's Program on International Policy Attitudes found. Of the nearly 15,000 people surveyed across fifteen countries, only 22 percent thought that enough oil would be found in the future for it to remain a primary energy source.[92] According to Google data, the world "oil" was the sixth most commonly searched term related to economics in 2008. Only those terms having to do with the unfolding credit crisis drew more interest.[93] On Capitol Hill, U.S. lawmakers took turns grilling the heads of the country's major oil companies over the price of gasoline. But the proceedings were unable to identify any culprits on which to blame the extraordinary other than the familiar ones: supply and demand, set alight by uncertainty and fear.

It would often be the higher-ups of the Bush administration who were left in the unenviable position of having to explain away astronomical oil prices to a skeptical public. "There's no quick fix" for high oil prices, U.S. Treasury

Secretary Henry Paulson reiterated to reporters on a trip to Qatar in the summer of 2008. "I don't see a lot of short-term answers."[94] Unfortunately for Paulson—and as the rest of the country would soon find out, much to its regret—there *was* one quick fix for high oil prices, which was about to take effect. But it was an ugly one: economic collapse. It would soon cure high worldwide oil prices, if only temporarily, but it was a cure worse than the disease.

By late summer, it began to dawn on the market that the robust demand growth that had propelled prices higher over the last decade or so had simply vanished. An extraordinary credit crisis, the first strains of which had appeared almost a year earlier, was now fully underway. Early in the year, markets had thought there was a risk the economy would slow, but few had understood the gravity of the situation. By the time that Lehman Brothers collapsed in September, it was clear that the economy was in a tailspin. Investors rushed for safety, selling assets like oil whose prices were sure to fall if economic growth evaporated. The forward curve for oil flipped from a steep backwardation to a deep contango, with futures contracts far out into the future trading at an enormous premium to oil for immediate delivery, in a desperate signal for traders to put oil into storage as a glut ensued. Gravity had set back into the oil market, and with a vengeance. By August, prices closed as low as $112 on the NYMEX. OPEC members tried to slow the slide, belatedly trimming their output to varying degrees, but to little avail. Still more selling—$91 by September; $62 by October; $49 in November. They hit rock bottom at $33.87 the week before Christmas. It was a dramatic turn of events, even in a market famous for its booms and busts.

"The global market has been turned upside down since the summer," the International Energy Agency declared, to little controversy. Budgets for new exploration and production projects were slashed as executives tried to come to grips with a wildly uncertain price outlook. New refineries in countries like Saudi Arabia and India were put on hold indefinitely. Plans to develop deepwater oilfields off the coast of West Africa and Brazil were recalibrated. Still in the throes of the credit crisis, lending terms for risky exploration ventures tightened—and in some cases, particularly for renewable projects like biofuels that relied on high oil prices to pencil, dried up entirely. Drillers in the United States were pulling rigs out of production at a breakneck pace. "We're in remission right now," an executive with Royal Dutch Shell's exploration and production operation told the *New York Times*, understatedly, amid the collapse. The crash had had a "dampening effect" on energy production, but he doubted the decline would last. Give the economy a little time to recover, and "the energy challenge will come back with a vengeance."[95]

He was right.

The Quiet Revolution

The drama in the markets was more than enough action to keep the energy community mesmerized, but behind the scenes—and way off Wall Street—another, quieter revolution had been taking place. It would soon call into question everything that the world's energy experts thought they knew about the future of oil.

The revolution was almost imperceptible at first, at least to anyone outside the North American oil patch. It started with a man named George Mitchell, an intrepid independent natural gas producer, who had spent the better part of the 1980s and 1990s trying in every way he could think of to extract gas from the Barnett Shale in north-central Texas. Geologists had long known that shale—dense sedimentary rock located across large swaths of the United States—contained oil and gas trapped within its pores. The problem was that it was too expensive to get them out, at least using the available technology. But by the late 1990s, that had changed. Mitchell Energy had developed a way to use hydraulic fracturing (commonly known as "fracking") to get the gas out of shale. When Devon Energy acquired the shale operator in 2002, it combined its expertise in directional drilling with Mitchell's fracking capabilities. Put together, and energized by relatively high prices, directional drilling and fracking could unlock vastly more hydrocarbons in a shale seam than could be done using traditional techniques.

Soon, other independent producers were getting into the unconventional gas game, and not just in the Barnett. The oil and gas industry was sinking vast amounts of money into exploring for and production of oil and gas in North America. Surging prices for both goods led companies nearly to triple their outlays for drilling acreage, equipment, and manpower 2003 and 2008 to $150 billion.[96] From plays in Louisiana to Arkansas to Oklahoma, shale gas began to pour out of the ground.[97] By 2007, the boom had made its way to the Marcellus and Utica shale formation, a massive play stretching from New York to Ohio and Virginia. Two years later, with the U.S. economy in the doldrums, the gas market was more or less glutted, with the onrush of new supplies helping to keep prices below $5 per million British thermal units on the NYMEX. Energy experts were waking up to a shockingly promising reality: The United States was awash in natural gas.

But what did it mean for oil? The stunning developments in U.S. natural gas production went all but unremarked upon by oil analysts in the run-up to July 2008, despite the impressive results the so-called shale gale had started to chalk up. Yes, shale gas was a big deal, but there was no sign that the oil that could be extracted from shale would ever amount to more than a rounding error. Many experts believed that oil molecules were simply too large to flow through shale rock in large volumes.[98] There was certainly nothing in historical experience to suggest it could. Tight oil—the light sweet liquids that could be pulled out of shale—amounted to less than 150,000 barrels per day prior to 2009. Two short years later, though, the tide had turned.

With natural gas prices in the doldrums, U.S. drillers turned their attention to oil, which offered vastly more attractive returns. Producers had mothballed nearly half their rotary drilling rigs in the United States when oil and gas prices collapsed in late 2008. Six months later, they were rushing to put them back in operation. This time, though, it was not gas, shale, or otherwise, they were after. It was crude oil and other liquid fuel sources, like natural gas liquids. And it was starting to yield some big results. At first few noticed them. "Some interest drilling data has emerged from the American Petroleum Institute," wrote an understated blogger at the *Financial Times* in July 2010, one of the first in the major media to identify the upswing. A "surge in oil drilling" looked as though it was "really gaining traction" in wringing oil out of shale.[99] The gains were finally starting to show up in the data. For American oil production, it was a rebound that had been long in coming. Since 1986, the country had undergone a slow, painful decline in crude output, despite being among the most heavily drilled real estate in the world. Back then, U.S. crude oil output had topped out at just over 9 million barrels a day, then ebbed slowly to around 5 million by 2008. Then it started to move higher, and with startling speed. In 2009, production rose by nearly half a million barrels a day over the course of the year. For a country whose oilfields experts had all but unanimously believed to be in decline, this was no mean feat. The next year, crude supplies climbed another 150,000 barrels a day higher. U.S. oil was not alone. Canadian oil swelled by the same amount in 2010. A great reversal was underway.

A boom in U.S. and Canadian oil production was gaining steam. It came from several places. In western Canada, vast oil sands—a mixture of sand, clay, bitumen, and other minerals so dense it has to be mined rather than pumped—were being exploited in growing quantities. The resources had been deemed uneconomic by oil companies a decade and a half earlier, but thanks to better technology and higher prices, that was no longer true. In the United States, production of so-called "tight oil" was soaring. Like gas, the combination of fracking and directional drilling was proving a potent duo, just as it had multiplied the country's natural gas production several years earlier. Starting around 2008, tight oil production took off like a rocket. Much of the growth happened in North Dakota's Bakken and South Texas' Eagle Ford shale, which catapulted domestic tight oil output to nearly 1 million barrels per day by 2011.[100] The frenetic pace of drilling activity all but entirely spared the towns that lay above these shale plays from the unemployment crisis in the rest of the country. Truckers, roughnecks, hoteliers, construction workers and the rest of the cast of characters needed to make shale happen in places like North Dakota could not be hired quickly enough. In other parts of the world, other oil discoveries were also starting to make headlines. Giant oil fields were located off the coast of Brazil and Africa. Iraqi production had started to gain traction, and the countries of the former Soviet bloc were briskly ramping up their output. All of a sudden, a landscape far different from shortage-obsessed one that had launched oil prices to $147 in July 2008 was starting to come into focus. "Simply put," wrote a *New York Times* reporter, "the world of energy has once again been turned upside down."[101]

Propelling this wave of oil onto the market was the return of extraordinarily high oil prices. It was only a matter of months before crude oil prices had bottomed out in the spring of 2009 that they snapped back beyond $60 per barrel yet again. A year on, WTI crude was hovering around $80 per barrel. Although the economic collapse had decimated oil demand in the developed world, in the past decade's major engines of demand growth—China, the Middle East, and elsewhere—people were still buying. The fact that some of these economies were spared the worst of the slowdown was partly why; but other factors, like massive fuel subsidies in much of the world, were also at play. Not only was this North America-led supply boom failing to cause a collapse in the global price of crude oil, it was a boom that would not have come to pass without high prices. Had WTI crude stayed at $40 per barrel in the years following the financial crisis, the American tight oil bonanza would have been stopped in its tracks. By the beginning of 2011, with most of the world's high-quality crude oils selling at in excess of $100 per barrel, unconventional oil production was thriving. Yet the growing volumes of North American liquids in pipelines, railcars, and barges were only enough to offset largely disappointing production results elsewhere in the world, rather than bring about a speedy, final collapse to oil prices—much to the regret of U.S. drivers. Perhaps more than at any other time in history, the U.S. oil economy was one of staggering abundance and simultaneous scarcity. Oil production in the United States would soon begin to grow at a faster pace than anywhere else in the world, totally defying the most optimistic predictions of a few years earlier, even as annual WTI crude prices, adjusted for inflation, notched their highest levels in the history of the industry.

"What If We Never Run out of Oil?"

By 2011, the unfolding revolution was making waves in the global market. The flood of Canadian and U.S. crude and natural gas liquids (a group of ultra-light hydrocarbons including ethane, propane, and butane) was pouring into storage facilities in the U.S. Midcontinent so quickly that there was hardly room to contain it. As storage there filled, the price of WTI crude, the benchmark American crude on which NYMEX futures were priced, started to fall further and further below the price of other crudes traded on the world market. Cushing, Oklahoma, where WTI was priced, was glutted. There simply was not enough space in pipelines, railcars, trucks, and barges to get the stuff to the U.S. Gulf Coast, where it could be refined and exported. Traditionally, WTI had sold at a premium of a few dollars to Brent crude oil, the benchmark crude in Northwest Europe, and the physical basis for the other major crude futures contract. But as Cushing filled, WTI prices started to sell at a discount, the market's way of providing a discount on the crude to spur demand, tame supplies, and relieve the overhang. Over the course of the year, WTI prices fell further and further below Brent prices. By late autumn, the spread had blown out to almost $30 per barrel. A decade prior, oil experts reasonably feared that the WTI benchmark might not survive, believing that oil output in the Midwestern United States might dry up so quickly as to distort pricing. Now the reverse had come true. It was overabundance, not scarcity, that was throwing the inland U.S. crude market into disarray.

While Wall Street worried about WTI-Brent, geopolitical thinks and policy analysts were scratching their heads about what the enormous reversal in American oil and gas fortunes meant for power politics. Five years prior, policymaker conferences and academic workshops had pondered whether looming energy scarcity would mean more and more wars fought over oil and gas supplies, or economic collapse driven by endlessly rising fossil fuel prices. Now the questions of the day were far different. Would North America soon replace OPEC as the 800-pound gorilla in the oil market? Was energy abundance the new order of the day? A growing chorus of experts thought so.

"Adios, OPEC," wrote Amy Myers Jaffe in *Foreign Policy*. "For half a century, the global energy supply's center of gravity has been the Middle East. This fact has had self-evidently enormous implications for the world we live in—and it's about to change." Astounding volumes of unconventional oil and gas—challenging offshore deposits, shale deposits, oil sands, and other heavy oil—were poised to come on stream in growing quantities. "Peak oil? Not anytime soon."[102] Yergin, writing the *Wall Street Journal*'s Saturday essay in September 2011, agreed: Peak oil, however enrapturing five years earlier, was on its way out the window. "Hubbert's Peak is still not in sight," he explained, citing the dramatic increases in estimates of the world's total resource endowment and likely production over the coming decades. Yes, the world has produced around 1 trillion barrels of oil since the origins of the industry well over a century prior. Yet at least 5 trillion barrels more are likely underground, around a third of which is already accessible given today's prices and technology, counting proved and probable reserves.[103] The end of the oil age was not nigh.

So much new oil and gas in the United States were enough to revive talk about a starry-eyed goal that had eluded U.S. presidents since the days of Richard Nixon: Energy independence. Ever since the oil crises of the 1970s, American politicians—not to mention drivers—had yearned for a return to the good old days of cheap, reliable supplies. Imports from the Middle East and elsewhere conjured up an image of unhealthy, even dangerous, dependence. Now, with U.S. oil and gas fields experiencing an unexpected renaissance, the dream of energy independence was rekindled as net oil imports declined. New supplies were not the only reason why—so was the fact the U.S. demand was in decline. In 2005, the country's oil consumption hit an all-time high, averaging 20.8 million barrels a day. Then it started to fall, declining in all but one of the next seven years. Exceptionally high fuel prices were pushing consumers to consume less, often by reducing the amount of time they spent behind the wheel. The Great Recession made the hurt of high prices all the worse. The country's cars and trucks were also becoming more efficient, able to run farther on less oil. To top it off, an aging population meant that more of the nation's drivers simply did not drive as much as they once did. Added together, rising supplies of oil and biofuels, coupled with declining demand, meant the United States was importing less oil. The United States imported 60 percent of its oil in 2005; seven years later, that figure had dropped to 41 percent. President Obama, in his 2013 State of the Union Address, celebrated the progress toward U.S. energy independence. "After years of talking about it," he triumphantly declared, "we're finally poised to control our own energy future."[104]

In the infancy of the American energy boom, experts disagreed sharply over whether the new sources of production would be sizeable enough to put peak oil to rest. Yes, U.S. and Canadian oil might be booming, but the gains in that part of the world might only offset declines in other places. This might just be another case of overexcitement, and a global production peak was still right around the bend after all.

By 2013 or so, though, there was nearly universal acceptance among the energy community that a tectonic supply shift was underway. Even analysts at the IEA, known for being circumspect in their pronouncements, were spreading the new energy gospel. "North American supply is an even bigger deal than we thought. A real 'game changer' in every way," the Agency's Executive Director, Maria van der Hoeven, told an industry conference in London in May 2013. "Not just because of the volumetric growth involved, but for a host of compounding reasons. . . . North America has set off a supply shock that is sending ripples throughout the world." It was an American boom, but it was headed worldwide in the decades ahead, "potentially leading to a broad reassessment of reserves." A generational realignment of oil production, consumption, trade, refining, and storage was taking place, and the rate of change would only grow. "There is hardly any aspect of the global oil supply chain that will not undergo some measure of transformation over the next five years, with significant consequences for the global economy and oil security," the IEA predicted.[105]

What of peak oil? Its popularity had come full circle. The alarming idea had crept from the shadows in the late 1990s to the limelight during the great bull market of 2003 to 2008, only to move way off Broadway five years later. When the term did get a mention in any mainstream media, it was hardly ever kind. "Has peak oil gone the way of the Flat Earth Society?" one news headline asked in March 2013. The idea had indeed fallen on hard times. "It's seen better days," Charles Ebinger of the Brookings Institution opined. "But I still go to conferences where I occasionally hear people arguing it." The consensus view, with the benefit of 20/20 hindsight, was that the theory behind peak oil was not built on a solid foundation. "I don't think it was ever a well-founded theory," one UCLA professor put it.[106] Whether more oil and gas was a good thing for the environment or climate was still a matter of sharp debate—but few people were still arguing that supplies would dry up any time soon. "New technology and a little-known energy source suggest that fossil fuels may not be finite," read the cover story of the *Atlantic* magazine in May 2013. "What if we never run out of oil?"[107]

It was not the first time in American history that someone had asked that question.

6

Conclusion

Rumors of oil's imminent demise have always proven to be exaggerated. Yet the stories behind these rumors, in their heyday, can be a potent force in the national conversation about the future of energy and in the marketplace about oil prices. No understanding of the oil market is complete without accounting for these episodes of fear about an impending, irreversible shortage of oil, accompanied by widespread visions of ever-rising prices. As far back as the dawn of the twentieth century, this cycle of hope and pessimism has influenced how people think about oil, from the street corner to the Oval Office. Years of feast and years of famine have each taken their turn in the limelight, one eventually giving way to the other. Lean times, marked by high or rising prices and tightening market conditions, have sometimes held sway for years. But the end of oil—an unyielding stagnation in production heralding the dawn of a new, quasi-permanent bull market for black gold—has never come about in the way that pessimistic prophets have envisioned. There is no telling what the future might hold when it comes to oil production and oil prices. They might surprise us in unforeseeable ways. Yet recognizing the patterns of the past provides an invaluable touchstone as we try to peer into the future.

The Drivers and Dynamics of New Era Anxiety in the Oil Market

What accounts for these chronic bouts of fear over an imminent shortage of oil, despite their having been proven wrong, time and again? The same kind of new era economic thinking that Yale economist Robert Shiller has identified in stock market and housing market booms can also drive public opinion about oil. A shift in market fundamentals—perhaps a loss of spare production capacity due to swelling demand—begins to push prices higher. People misinterpret rising prices as signaling that a new era, perhaps of permanent shortage, has appeared. They predict that oil production has reached an unbreakable ceiling, which will mean an end to rising oil supplies. The newfound scarcity will cause prices to continue to rise far out into the future, and perhaps perpetually, they argue. They fail to see the parallels between the latest bout of worry and prior ones that have come and gone over the course of history, as well as to remember the industry's cyclical nature. Experts make bold prophecies that draw widespread attention and cater to a press hungry for riveting, notable stories. People accept exaggerated stories about the implications of these new developments on the future of oil decades or even centuries hence. But when oil production continues to grow, demand cools, and prices moderate, the once-popular new era claims fall out of fashion.

The idea that the world is running out of oil becomes contagious when roaring demand pushes prices higher for long periods of time. The shortage idea gains popularity as prices rise. It falls out of fashion as prices fall. The source of the increase in prices also matters. Shortage-driven fears tend to arise when swelling consumption—rather than a production-cutoff or other supply interruption—is at least partly behind the rise in prices. The longer and steeper the rise in prices, the better the odds that the notion of a world out of oil will gain currency. A momentary jump in prices is not nearly as likely to spawn widespread talk of a worsening shortage of oil as a prolonged period of ever-increasing consumption. The fascinating part of this pattern is that more and more oil is being used up every year—but people only seem to think we're "running out" of the stuff when prices are rising. Once they fall, as they inevitably do, all appears well again, despite the fact that the world may be gulping down more and more of the black gold.

By the same token, however, a dangerous mirror-image of the same thinking can take hold when oil production is rising and prices are low. The cynics are not the only ones who can fall prey to new era naïveté. The notion that the world will be awash in cheap oil in perpetuity can become problematically popular when supply exceeds demand for a long period of time, only to be swept away when the market tightens again and prices begin to rise. Optimism risks becoming Pollyanna-ish when, in the good times, people see an energy feast that they doubt will ever turn into famine. That line of reasoning has always proven foolhardy. Complacency among oil-consuming countries, like the United States, can set in all too easily in times like the late 1980s or again in the late 1990s, in the wake of an oil price collapse.

Shortage anxiety in the oil market has some special attributes not apparent in other markets. When it comes to the stock market, for instance, bull markets are always associated with unbridled enthusiasm—the outlook is rosy. Not so for oil. The cultural and political moments that fuel bouts of worry that the world's fossil fuels will soon be gone are often overwhelmingly gloomy or fretful. In oil, unlike in equities, bleak predictions about geology and geopolitics are the stuff of bull markets. Furthermore, when shortage worry strikes in the oil market, forgotten voices from the past who had claimed to foresee the end of oil supplies and a looming price spike are brought center stage and heralded as prophets. Moreover, some people who dislike oil as an energy source seem all too eager to fan the flames. As far as they are concerned, it might not be such a bad thing if oil did run out after all. In their eagerness to move society away from oil-based technologies, they can become highly receptive to messages from experts about a looming end to oil, which they then pass along to others. Those who subscribe to the shortage idea chronically underestimate the power of future innovation to unlock new sources of supply, as well as the power of high prices to moderate demand. But in oil, as in life, such static assumptions never seem to play out.

Lessons for Wall Street and Washington

What lessons for today do the last hundred years of history hold? What can yesterday's shortage scares teach those who navigate these markets, whether as investors or as public officials who regulate the oil industry and shape the geopolitics of oil? Here are four lessons.

One of the most profound lessons of the long sweep of history is also the most simple: Never underestimate the power of the market to wring abundance out of scarcity. Markets—even one as imperfect as oil, where small groups of powerful suppliers have held enormous sway at turns over the course of history—work astonishingly well when it comes to increasing supplies. Supply, demand, inventories, and price: the components of the oil market are simple, perhaps deceptively so. Their elegant interplay, whereby a jump in one—say, prices—can work wonders in another—in this case, supply—is profoundly powerful. The invisible hand capable of pushing along human progress, which the legendary economist Adam Smith identified centuries ago, is still capable of pushing mankind along in its search for material well-being—even when the odds appear hopelessly long.

Yet the picture is not entirely rosy. Oil crises will come in the future, as they have in the past. Yes, markets may work wonders in increasing the abundance of scarce materials—but they often do so slowly, at a considerable lag, and require painfully high prices. A surge in prices to a new high is typically a prerequisite for coaxing more oil out of the ground, following the simple laws of supply and demand. History is littered with overly optimistic economists who incorrectly assumed that the price of oil in the future would look very similar to the price of oil in the past because any drift upward would be quickly self-correcting. The lesson of history—in this case, recent history—is that it can take upward of a decade, if not longer, for rising prices to stimulate the innovation and risk-taking necessary to unlock new resources at scale. In the bull market of the 2000s, for example, some analysts cast doubt on the idea that crude oil prices could shoot beyond $25 to $30 per barrel for an extended period. Even analysts who recognize the surprising power of market dynamics to increase supplies of a resource like oil should avoid assuming that the oil will come online smoothly or immediately—or that prices will be low.

Those who bet against the forward march of technological innovation have been on the wrong side of history, time and again. Today's unconventional sources of oil have always become tomorrow's conventional ones. In their search to extract ever more oil from the ground, every generation has viewed its own challenges as hopelessly more complex than what their predecessors had to face, and perhaps even unsolvable. They fail to realize that every age's problems seem overwhelming at the time. But surging prices have always brought forth unforeseen bounty, defying the skeptics. As Dylan Grice of Société Générale memorably put it, "When you buy commodities, you're selling human ingenuity."[1] Shorting mankind's ability to innovate its ways out of perceived scarcity has always proven a dangerous proposition.

All of this is not to say that policy does not have a role to play in boosting energy supplies. There is no question that sound policy can help make markets prosper, providing incentives that accelerate technological progress across the spectrum of energy sources and catalyzing their adoption by the private sector. America's current oil and gas boom, while proving the resilience of a well-greased private sector, also demonstrates public policy's potential for making meaningful contributions in the areas of primary research and development and via corporate tax breaks. Over the course of three decades, and working in close cooperation with industry leaders like George Mitchell, Washington invested in and nurtured cost-effective shale gas extraction technology in several ways, as a 2012 report by the Breakthrough Institute details. As early as the 1970s, the Eastern Gas Shales Project brought together corporate players and policymakers in a series of demonstration projects. The Gas Research Institute, an industry research initiative, was partially federally supported. Federal agencies, including the forerunners of the Department of Energy and the National Energy Technology Laboratory, pushed early shale fracking and directional drilling technologies along. Tax policy also played a role. The Section 29 production tax credit also supported shale development from 1980 through 2002.[2] Absent these initiatives, the boom might well have taken place later or more slowly.

The second lesson of history is that the oil trade is marked by disjunctures and discontinuity. Dramatic, unexpected changes over time are the rule, not the exception. The market's evolution moves forward in fits and starts, not smoothly or predictably, never standing still for years on end. This pattern is as true for supply and demand as it is for prices (not to mention for issues as basic as how oil prices are determined, which have evolved over time). Those who make predictions about the future based on the oil market as a static system, where little changes from decade to decade, have always missed the mark. Smart people in every era have succumbed to the folly of believing that contemporary market trends—whether in terms of supply, demand, or prices— would stay mostly unchanged over the next few decades. They never have. In fact, it can actually be those moments of apparent tranquility—when oil prices are low or high for long periods of time—that lay the foundation for massive upheaval. Policymakers, investors, and market analysts should beware of putting too much trust in long-term predictions about the oil market that do not acknowledge the high degree of uncertainty that shrouds them. Public officials should be extremely cautious when making forward-looking statements about oil prices and supplies.

The oil market, like that of other commodities, is cyclical. New era thinkers throughout history have always overlooked this critical point. Low prices pave the way for high ones, only to have high prices sow the seeds of low ones, and so on. Few people have explained the cyclical nature of the market better than Matthew Simmons did in 2000, before he had become a dogged defender of the peak oil idea, in describing how cycles take shape in natural resource markets. When it comes to "cycles of commodities," he explained, "the patterns are always the same":

> Demand for a particular crop ends up growing too fast. Supplies then get short and the price soars. The farmer quickens his planting cycle to capture these high prices just as demand is starting to fall due to being too high. This creates a larger glut. Prices then plunge. The farmer stops planting. As supplies then dwindle, low prices begin to stimulate demand. As a result, commodities swing back and forth, rocketing from peak to bottom and back to peak.

"The only real difference between agriculture and energy," Simmons went on to explain, "is that the cycles are longer" in oil. "Ten years from now, all you guys will be discussing the likelihood of $200 oil just as demand is dropping and supplies are on the rise!"[3] He was right. So right, in fact, that it appears he might have lost sight of just how long seasons of flat production can last when he became an outspoken proponent of peak oil in the mid-2000s. The dynamic between prices, supply, and demand is cyclical, in the truest sense: It is not just that feast gives way to famine, but feast actually creates the conditions that make famine inevitable. True, there are stabilizing forces in the market and periods when oil prices (or their trajectory) stay remarkably constant. The steadying hand of some political force capable of dictating production, thereby balancing the market, can contribute to these periods of unusual calm. The Texas Railroad Commission and OPEC are two examples from history. Yet such tranquility can only partially mask the boom-and-bust nature of the oil market, whose exceptional volatility is coded in the very economic DNA of commodities like oil.

It is essential to acknowledge the always-evolving nature of the oil market to see the danger in naively extrapolating the recent past too far into the future. All too often, waves of fear about a world supposedly running out of oil have been due in part to a bias that behavioral economists call the representativeness heuristic: After finding a trend or pattern in a relatively small sample of data, they erroneously extrapolate the trend far out into the future. People deciding which mutual funds to pick for their 401(k), for instance, will place too much emphasis on which fared the best over the last year or two, assuming that solid trajectory would keep on going. Such assumptions are dangerous. Over-extrapolating the past far out into the future leaves little room for tomorrow's trends to be different than today's. It naively assumes that things will stay about the same (or at least keep changing in the same way it has in the recent past), when in fact, over the course of history, assuming continuity has missed the mark. New era thinkers get into trouble when they project today's market trends far out into the future.

Public officials should be modest in their guesses about what is around the bend for the energy industry. Erroneous statements by government experts, whether within the U.S. Geological Survey, the Oval Office, or the International Energy Agency, have fanned the flames of every shortage crisis. Oil prices are not at all easy to predict nor are oil reserves simple to quantify. Prices follow what statisticians call a "random walk" over long periods of time, according to a study by James Hamilton of the University of California-San Diego. In other words, prices one day, year, or decade out into the future cannot be reliably calculated based on past trends. The odds that your forecast will prove massively wrong grow quickly the farther out in the future you are trying to peer. Most people who have tried to predict oil prices over the last century have been far too slow to acknowledge just how prone to error such forecasts are.[4] Prices zig when people think they will zag, inevitably. Even numbers as seemingly unchanging as the size of the world's recoverable oil reserves fluctuate massively over time, as changes in prices, politics, and technology add to or diminish how much oil can be pumped at any given time. Each successive generation has believed that they were living through the "end of history" in oil, naively assuming that price trends at that moment would hold constant far into the future. None was right, and there is little reason to believe that it will be different from here on out.

Well-advised public policy does not try to predict where energy markets are headed. Instead, it puts in place a set of rules under which a country's energy economy can thrive in many different environments. It aims for resilience, not prescience. It strives for an energy economy that is robust against a broad range of scenarios, even ones that appear unlikely at the time, whether for oil production, consumption, or prices. The same rule holds true for other energy sources beyond oil, such as natural gas and renewable forms of energy. If public officials can acknowledge how often the future of energy has surprised them (and their predecessors, who were at least as smart as they are), and craft policy accordingly, energy supplies are more likely to be plentiful and affordable over the long run.

What does this approach mean in practice? Such a paradigm embraces energy efficiency technologies. Although fuel-efficient cars, trucks, and appliances are less vital when energy prices are low, they provide a margin of safety for the time when prices will eventually bounce back, perhaps with a vengeance. The resilience-over-prescience paradigm favors practices that mitigate a company or country's vulnerability to sudden, unexpected changes in prices—in either direction, not just up. For centuries, sudden drops in natural resource prices have corresponded closely with sovereign default rates in countries, usually in the developing world, that depend on these exports for fiscal survival.[5] Far-sighted energy policy eschews propping up companies whose existence is only possible in isolated policy and price environments that are unlikely to be sustained. Trying to pick winners among early–stage energy ventures is as foolish as declaring with certainty where oil prices will be in ten years' time. Rather, public policy should provide targeted support, via tax breaks and other initiatives, in research and development that has commercial applications but may be too early stage or risky for businesses to tackle alone. Shale gas production, as explained earlier, was a successful case study in this kind of public-private partnership. A resilience-over-prescience policy paradigm eschews policies like the Powerplant and Industrial Fuel Use Act of 1978, which banned the use of natural gas for electricity generation, based on the Carter administration's view that the country's gas resources were near extinction and "too few" domestic utilities were burning coal, "our most abundant energy source."[6] Cutting natural gas demand off at the knees, however, caused prices to be much lower than they otherwise would have been, stifling innovation in that much-cleaner energy source and giving rise to a generation of coal-burning plants that today's officials are wringing their hands over.

As for a third lesson, it is this: Basic truths about how markets work—what makes supplies go up and what drives consumption down, for instance—have proven astoundingly reliable, time and again. New era naysayers who claim that the "old" laws of economic gravity no longer apply have usually found themselves on the wrong side of history. Prophecies about some apocalyptic rupture in world energy supplies have come and gone over the last century. These intellectual fads have all shared a common element: They distrust the power of basic economic forces to yield the same results they have in the past. In so doing, they often commit several fatal errors. They fail to account for the fact that a prolonged, earth-shattering rise in prices would cause the amount of recoverable oil under the Earth's surface to climb dramatically, slowly call forth enormous new amounts of oil and simultaneously force consumers to cut back on their consumption, perhaps through a painful recession. Prophecies that inflation-adjusted oil prices will rise without ceasing often gloss over the fact that other market forces will work against such an outcome over long periods of time. Their lack of confidence in simple market mechanisms to operate has always proven groundless.

When shortage fevers are at their peak, the temptation to lose confidence that oil production might increase can beguile all but the most disciplined analysts. Why? For one thing, it is because no one at the time can see *where* more oil would come from. People massively overestimate how much they know about the world's oil reserves. They rashly conclude that if it has not been found yet, or cannot be extracted with today's technology or at today's prices, that it likely will not ever be. Many of those pumping oil in Pennsylvania in the mid-nineteenth century fancied that there might not be any oil in the Americas outside of Appalachia. Later, in the 1920s, Americans were once again stunned to discover just how much oil lay beneath Texas, Oklahoma, California, and elsewhere in the southwestern United States, not to mention abroad. The almost incomprehensible size of Middle Eastern oil reserves shocked U.S. officials—again—in the mid-twentieth century. Within the last few years, the vast amounts of tight oil recoverable in North America and elsewhere yet again upended what the world thought it knew about energy supplies.

Each time, people have had trouble seeing *how* the world could possibly produce more oil. In the moment, the supply situation really *did* look different to contemporary observers—until, all of a sudden, it didn't (again). Investors who have trust in the simple economic principle that high prices do spur growth in supplies over time, thanks to technological innovation, inoculate themselves against new era shortage fever.

The fourth lesson of history is that the best defense against unfounded worries is good data. Scares thrive when information is scarce. Wrongheaded predictions abound when public access to timely, accurate, comprehensive estimates of global oil reserves, production, and consumption is wanting. Conspiracy theories and confusion take hold more easily when energy markets are opaque. Whether in Washington or Riyadh, Beijing or Caracas, if policymakers are serious about making oil prices more stable and predictable, they will invest in tracking and disclosing data about fundamental conditions in the oil market, making sure information is reliable and readily available. The U.S. Energy Information Administration, which provides the country's official energy statistics, does an excellent job in this respect, and serves as a role model for similar bodies around the world. These measures will not be a cure all. Better data will not always mean good enough data. Wrongheaded conjecture and paradigmatic misunderstanding of market behavior is an inescapable feature of how markets work in the real world. Yet better data increases the odds of decisions by market participants that reflect actual conditions in the marketplace and draw on all known information about what the future might hold for supply, demand, and prices.

Improving transparency, within both the physical and financial oil markets, holds the largest potential for decreasing unwarranted price volatility and making certain that retail prices reflect economic fundamentals.[7] A well-functioning physical oil market, where price discovery occurs efficiently and stably, requires the transparent flow of accurate information about supply and demand. Yet, as anyone who has ever closely studied the oil market knows, data on oil supply and demand is limited at best and often contradictory. Unlike more transparent markets, like the U.S. stock market, fundamental conditions in the global oil trade are often difficult to discern, with limited data often forcing analysts to rely on best guesses. Only those nations within the OECD reliably release detailed information about the flow of oil within their own borders. The U.S. Energy Information Administration sets the standard in this respect, providing high-quality data and analysis about market conditions within the United States and the international market as well. Yet the problem is not one any country can fix by itself. Because the oil market is global, what goes on in one corner of the world spills echoes across the rest of the world.

Likewise, greater financial transparency could help ensure that financial market prices reflect fundamental conditions as closely as possible. The CFTC should expand its Commitment of Traders report, which details buying and selling activity on a weekly basis according to trader categories. These categories should be more narrowly defined. The report might also include information about where these various classes of traders are taking positions along the forward curve. This added detail about market activity organized by term structure could help regulators and market participants identify where in the market risk might be multiplying, especially in less liquid contracts.

Of all of these regulatory options, the ones with the clearest benefit and lowest risk focus on improving transparency, particularly in the physical oil market. And yet they are also likely the hardest to achieve. Scrubbing away the opacity in the physical market would require a host of major oil suppliers and consumers to voluntarily disclose their dealings. Western nations do not have the regulatory reach to command such compliance. Yet there are steps that could be taken. In terms of data on oil consumption, the United States and other OECD countries would be well advised to find ways to entice India and China into the IEA fold. Their oil demand growth through 2030 will represent half of the worldwide total. Even if IEA membership for them is not in the cards—after it, China would likely bristle at the data reporting requirements membership would entail—providing incentives for both countries to sacrifice resources and privacy for the good of oil price stability should be a point of foreign policy emphasis by IEA members. After all, these two growing consumers both share the OECD's interest in a well-functioning oil market.

Data from critical oil suppliers on their reserves, production, and exports is becoming better though still incomplete. The Joint Oil Data Initiative (JODI), created in 2001, is a voluntary data reporting system that includes more than seventy nations that provide data on their domestic oil production, demand, and inventories. JODI has been a significant step in the right direction, but it is not enough: its numbers are still often at odds with other official sources and the incentive that OPEC members have to tailor their data to their own advantage is strong. A lack of credible public information about global oil reserves is also a problem. Much about oil reserves in OPEC countries is still unknown, a veil of secrecy that is a source of long-term price uncertainty.

The United States and its partners should pursue those regulatory initiatives within their jurisdiction even as they attempt to make strides in other areas. Physical transparency in the oil market would make the single largest contribution to market improvement, and yet it is the least amenable to dictation from the IEA. The United States, the United Kingdom, and other countries that are home to sizable financial sector activity should review their domestic regulations in these areas while seeking to minimize capital flight and overly stringent measures. They should couple their attempts to improve physical oil market transparency beyond their borders with broader foreign policy objectives where they may have more leverage. Although they will face headwinds from some major consuming and producing nations, these efforts can yield dividends over time by slowly removing the largest impediment to a more efficient oil market.

A New Era, or Just a Forgotten One?

"There is nothing new but that which has been forgotten," Rose Bertin once observed. Over the last century, countless aspects of the way mankind powers itself have evolved. Technology has improved, resources have been discovered, and civilizations have come and gone. Yet echoes of the past are with us today. Expert claims of an imminent, irreversible shortage of oil when prices are rising—or on the flip side, commentators' claims of limitless abundance when prices are falling—have been an enduring feature of the oil market over time, right up to the present day. Talk of a new era of scarcity (or abundance) can take in a broad swath of the public when prices are rising (or falling), fueled by the latest research by prominent experts and government officials. The notion of oil being used up as production rises, leading to higher and higher prices over time, has a certain intuitive appeal. But visions of perpetually rising prices amid an exhausting resource have never quite been borne out the way their seers envision. The allure of profit pushes technology ahead, opens up new vistas of exploration, and eventually reassures the fretful that the end is still somewhere over the horizon. Bull market gives way to bear. The "end of oil" comes and goes—only to come again years later.

There is no telling what the future of energy holds. Yet those who look ahead and see the same conditions that exist in today's energy markets existing far out in the future stand on shaky ground. As for oil, it is all but certain that it will eventually cease to play an important role in the global energy mix. Some better source will supplant it. It is only a matter of time. But those who prophesy the commodity's demise would do well to remember the admonition of one of Riyadh's wise men on the future of fossil fuels: "The Stone Age came to an end, not because we had a lack of stones, and the oil age will come to an end not because we have a lack of oil."[8]

NOTES

Prelims

1. Daniel Yergin and Joseph Stanislaw, "How OPEC Lost Control of Oil," *Time* 151, no. 13 (April 6, 1998), 58.
2. See Google Books NGram Viewer, http://books.google.com/ngrams.
3. Oil prices taken from BP, *BP Statistical Review of World Energy June 2012* (London: 2012), http://www.bp.com/statisticalreview.
4. With a smoothing of 1. Google Books NGram data does not extend beyond 2008, unfortunately. The data end the very year that crude prices roared to an all-time high in nominal terms. Although crude remains expensive in nominal and real terms, media attention to the claim that the world's oil production may be hitting its permanent limit seems to have waned considerably. This loss of public attention has seemed especially pronounced since 2012 or so, when the notion of an American oil and gas boom began to spread.
5. Oil prices taken from BP, *BP Statistical Review of World Energy June 2012*. ProQuest search conducted January 2014 on the terms "run out of oil" and "running out of oil." The texts searched included newspapers, magazines, trade journals, scholarly journals, dissertations and theses, and other reports.
6. Bassam Fattouh, "An Anatomy of the Crude Oil Pricing System," WPM 40, Oxford Institute for Energy Studies, January 2011.
7. Colin J. Campbell, "About Peak Oil: Understanding Peak Oil," APSO International, http://www.peakoil.net.
8. Robert L. Hirsch, Roger Bezdek, and Robert Wendling, *Peaking of World Oil Production: Impacts, Mitigation, & Risk Management*, Report prepared by Science Applications International Corporation for the National Energy Technology Laboratory, U.S. Department of Energy (February 2005), 19.
9. Eyal Dvir and Kenneth S. Rogoff, "Three Epochs of Oil," NBER Working Paper No. 14927, National Bureau of Economic Research (April 2009).
10. Carmen M. Reinhart and Kenneth S. Rogoff, *This Time is Different: Eight Centuries of Financial Folly* (Princeton, NJ: Princeton University Press, 2009), xxv.
11. James D. Hamilton, "Cause and Consequences of the Oil Shock of 2007–08," Brookings Papers on Economic Activity (Spring 2009); James L. Smith, "World Oil: Market or Mayhem?" *Journal of Economic Perspectives* 23, no. 3 (Summer 2009), 145–164.
12. Michael Levi, *The Power Surge: Energy, Opportunity, and the Battle for America's Future* (New York: Oxford University Press, 2013).

Chapter 01

1. Brian Black, *Petrolia: The Landscape of America's First Oil Boom* (Baltimore, MD: Johns Hopkins University Press, 2000).

2. Andrew Carnegie, *The Autobiography of Andrew Carnegie* (New York: PublicAffairs, 2011), 136–141.

3. David Nasaw, *Andrew Carnegie* (New York: Penguin Press, 2006), 76–78.

4. Andrew Carnegie, *The Autobiography of Andrew Carnegie and the Gospel of Wealth* (Digireads.com Publishing, 2009), 69–70.

5. "Online Data Robert Shiller," Homepage of Robert J. Shiller. Accessed July 2014 at http://www.econ.yale.edu/~shiller/data.htm; and *BP Statistical Review of World Energy June 2014* (London: BP, 2014).

6. Glen Hiemstra, "Matt Simmons Sees $300 Oil," Futurist.com, March 10, 2008, http://www.futurist.com/2008/03/10/matt-simmons-sees-300-oil/.

7. Mimi Swartz, "The Gospel According to Matthew," *Texas Monthly*, February 2008.

8. Colin J. Campbell, "Understanding Peak Oil," http://www.peakoil.net/about-peak-oil. Although the website does not have a date of publication, Tom Bower in *Oil: Money, Politics, and Power in the 21st Century* puts it at June 1996.

9. Colin J. Campbell, *The Golden Century of Oil 1950–2050: The Depletion of a Resource* (New York: Springer, 1991), pp. ix–x, 51–52.

10. Kenneth S. Deffeyes, *Hubbert's Peak: The Impending World Oil Shortage* (Princeton: Princeton University Press, 2001), 1.

11. See staff piece, "Responses to Daniel Yergin's Attack on Peak Oil," September 19, 2011, http://www.resilience.org.

12. WorldPublicOpinion.org, "World Publics Say Oil Needs to be Replaced as Energy Source," World Public Opinion.org, Press Release, April 20, 2008, http://www.worldpublicopinion.org/.

13. See Steve Andrews and Randy Udall, "Peak Oil: It's the Flows, Stupid," May 12, 2008, http://www.resilience.org.

14. Daniel Yergin, "It's Not the End of the Oil Age," *Washington Post*, July 31, 2005.

15. For three excellent overviews of this debate, see John E. Tilton, *On Borrowed Time?: Assessing the Threat of Mineral Depletion* (Washington, DC: Resources for the Future, 2003); Eric Neumayer, "Scarce or Abundant? The Economics of Natural Resource Availability," *Journal of Economic Surveys* 14, no. 3 (2000), 307–335; and George A. Nooten, "Sustainable Development and Nonrenewable Resources—A Multilateral Perspective," in *Proceedings of the Workshop on Deposit Modeling, Mineral Resource Assessment, and Sustainable Development* (2007).

16. Thomas Robert Malthus, *An Essay on the Principle of Population*, 6th ed. (London: John Murray, 1826).

17. David Ricardo, *On the Principles of Political Economy and Taxation*, 3rd ed. (London: John Murray, 1821).

18. John Stuart Mill, *Principles of Political Economy: With Some of their Applications to Social Philosophy*, 5th ed. (London: Parker, Son, and Bourn, 1862).

19. William Stanley Jevons, *The Coal Question*, 2nd ed. (London: Macmillian and Co., 1866).

20. Harold Hotelling, "The Economics of Exhaustible Resources," *Journal of Political Economy* 39, no. 2 (April 1931), 137–175.

21. For a useful discussion of the Hotelling rule, see Tobias Kronenberg, "Should We Worry about the Failure of the Hotelling Rule?" Paper prepared for Monte Verità Conference on Sustainable Resource Use and Economic Dynamics, June 4–9, 2006, Ascona, Switzerland. Center of Economic Research at ETH Zurich.

22. A minority of economists would argue otherwise. For a discussion, see P. G. Bradley and G. C. Watkins, "Detecting Resource Scarcity: The Case of Petroleum," in *Proceedings of the IAEE 17th International Energy Conference, Stavanger, Norway, May 25–27*, vol. 2 (1994); M. A. Adelman and G. C. Watkins, "Reserve Asset Values and the Hotelling Valuation Principle: Further Evidence," *Southern Economic Journal* 61, no. 3 (January 1995), 664–673; James D. Hamilton, "Oil Prices, Exhaustible Resources, and Economic

Growth," in Roger Fouquet (ed.), *Handbook on Energy and Climate Change* (Cheltenham, UK, and Northampton, MA: Edward Elgar Publishing, 2013).

23. Hamilton, "Oil Prices, Exhaustible Resources, and Economic Growth."

24. Donella H. Meadows, Dennis L. Meadows, Jorgen Randers, and William W. Behrens III, *Limits to Growth* (New York: Universe Books, 1972).

25. Herman E. Daly, "Steady-State Economics," in Matthew Alan Cahn and Rory O'Brien (eds.), *Thinking about the Environment: Readings on Politics, Property, and the Physical World* (New York: M. E. Sharp, 1996), 250–255; Nicholas Georgesçu-Roegen, *The Entropy Law and the Economic Process* (Cambridge, MA: Harvard University Press, 1971); Nicholas Georgesçu-Roegen, "Energy and Economic Myths," *Southern Economic Journal* 41, no. 3 (January 1975), 247–381; Ezra J. Mishan, "Growth and Antigrowth: What are the Issues?" in Andrew Weintraub, Eli Schwartz, and Jay Richard Aronson (eds.), *The Economic Growth Controversy* (London: Macmillan, 1974), 3–38.

26. See Wilfred Beckerman, "Economists, Scientists, and Environmental Catastrophe," *Oxford Economic Papers* 24, no. 3 (November 1972), 327–344; and Wilfred Beckerman, *In Defence of Economic Growth* (London: J. Cape, 1974).

27. William D. Nordhaus, "Resources as a Constraint on Growth," *American Economic Review* 64, no. 2 (May 1974), 22–26.

28. Joseph Stiglitz, "Growth with Exhaustible Natural Resources: Efficient and Optimal Growth Paths," *Review of Economic Studies* 41, Symposium on the Economics of Exhaustible Resources (1974), 123–137.

29. M. K. Hubbert, "Nuclear Energy and the Fossil Fuels," *Drilling and Production Practice*, American Petroleum Institute (1956).

30. Steven M. Gorelick, *Oil Panic and the Global Crisis: Predictions and Myths* (Oxford: Blackwell, 2009), 90–91.

31. Ibid., 92.

32. Ibid., 95. Also see U.S. EIA data on domestic natural gas gross withdrawals on a monthly basis, available through the EIA's online data portal., http://www.eia.gov/naturalgas/data.cfm

33. International Energy Agency, *Oil Market Report: 12 December 2012*, December 12, 2012.

34. Gorelick, *Oil Panic and the Global Crisis*, 95–97.

35. U.S. Geological Survey, *An Estimate of Undiscovered Conventional Oil and Gas Resources of the World 2012*, Fact Sheet 2012–2013 (March 2012), http://pubs.usgs.gov/fs/2012/3042/fs2012-3042.pdf.

36. Albert A. Bartlett, "Sustained Availability: A Management Program for Non-renewable Resources," *American Journal of Physics* 54, no. 5 (May 1986), 398–402; Albert A. Bartlett, "Reflections on Sustainability, Population Growth, and the Environment," *Population & Environment* 16, no. 1 (September 1994), 5–35; Colin J. Campbell, "The Coming Oil Crisis," *Quarterly Review of Economics and Finance* 42 (1997), 373–389; Colin J. Campbell, *Oil Crisis* (Essex: Multi-Science Publishing Co., 2005); John D. Edwards, "Crude Oil and Alternate Energy Production Forecasts for the Twenty-first Century: The End of the Hydrocarbon Era," *American Association of Petroleum Geologists Bulletin* 81, no. 8 (1997), 1292–1305; Colin J. Campbell and Jean H. Laherrere, "The End of Cheap Oil," *Scientific American*, March 1998.

37. M. A. Adelman, "Mineral Depletion, with Special Reference to Petroleum," *Review of Economics and Statistics* 72, no. 1 (February 1990), 1–10; M. A. Adelman, "The World Oil Market: Past and Future," *Energy Journal* 15, Special Issue (1994), 1–11; Richard L. Gordon, "IAEE Convention Speech: Energy, Exhaustion, Environmentalism, and Etatism," *Energy Journal* 15, no. 1 (1994), 1–16; P. J. McCabe, "Energy Resources—Cornucopia or Empty Barrel?" *American Association of Petroleum Geologists Bulletin* 11 (1998), 2110–2134; Gorelick, *Oil Panic and the Global Crisis*; Paul J. Stevens, "The Future Price of Crude Oil," *Middle East Economic Survey* 47, no. 37 (September 2004); Paul J. Stevens, "Oil Markets," *Oxford Review of Economic Policy* 21, no. 1 (2005), 19–42, http://web.archive.org/web/20041216111616/http://www.mees.com/postedarticles/oped/a47n37d01.htm

38. Julian L. Simon, "Resources, Population, Environment: An Over-supply of False Bad News," *Science* 208, no. 4451 (June 1980), 1431–1437; Julian L. Simon, "False Bad News is Truly Bad News," *The Public Interest*, no. 65 (1981), 80–90.

39. Paul Cashin, C. John McDermott, and Alasdair Scott, "Characteristics of the Current Commodity Boom," in *Global Economic Prospects 2009: Commodities at the Crossroads* (Washington, DC: The World Bank, 2009); David I. Harvey, Neil M. Kellard, Jakob B. Madsen, and Mark E. Wohar, "The Prebisch-Singer Hypothesis: Four Centuries of Evidence," *Review of Economics and Statistics* 92, no. 2 (May 2010), 367–377.

40. Raúl Prebisch, *The Economic Development of Latin America and Its Principal Problems* (New York: United Nations Department of Economic Affairs, 1950); Hans Singer, "The Distribution of Gains between Investing and Borrowing Countries," *American Economic Review* 40, no. 2 (1950). For a brief overview of the thesis, see "Prebisch-Singer Hypothesis," *International Encyclopedia of the Social Sciences* (2008), http://www.encyclopedia.com/doc/1G2-3045302026.html.

41. For details regarding the *Economist* commodity index, see "Markets & Data," *Economist*, http://www.economist.com/markets-data. For a good overview of how the index's component commodities and their weightings index have evolved since its creation, see "Appendix: The Economist's Industrial Commodity-Price Index" from Paul Cashin and C. John McDermott, "The Long-Run Behavior of Commodity Prices: Small Trends and Big Variability," *IMF Staff Papers* 49, no.2, International Monetary Fund (2002). For the so-called Harvey index, see Harvey et al., "The Prebisch-Singer Hypothesis," 367–377.

42. Shiller explains his methodology in compiling the historical CPI data set that starts in January 1871: "The CPI-U (Consumer Price Index-All Urban Consumers) published by the U.S. Bureau of Labor Statistics begins in 1913; for years before 1913 1 spliced to the CPI Warren and Pearson's price index, by multiplying it by the ratio of the indexes in January 1913. December 1999 and January 2000 values for the CPI are extrapolated." ("Online Data Robert Shiller," Homepage of Robert J. Shiller, http://www.econ.yale.edu/~shiller/data.htm.) I brought the data current to 2011 using recent Bureau of Labor Statistics figures.

43. Most studies of long-run trends in commodity prices make use of the Grilli-Yang (1988) index of raw materials prices. Rather than use that data set, though, I opt instead for the *Economist* industrials index, for the reasons outlined in Cashin and McDermott, "The Long-Run Behavior of Commodity Prices." As they note, the latter index includes four more decades' worth of data. Regardless, given the high degree of correlation between the two data sets—0.85 in Cashin and McDermott's extension of the Grilli-Yang data from 1900 to 1999—means that using either index should allow for substantially similar results.

44. For the current weightings, see "Economist Commodity Price Index: Weights in the Index," *Economist*, http://media.economist.com/media/pdf/Weights2005.pdf.

45. See the *Economist* industrial commodity price index since 1845, for example, in "Commodities: Crowded Out," *Economist*, September 24, 2011, http://www.economist.com/node/21528986, which is adjusted by the U.S. GDP deflator. As of the start of 2012, the index stands at roughly half its 1845 value in real terms. For a discussion of which deflator to use when assessing long-term commodity price trends, see John T. Cuddington, "Calculating Long-Term Trends in the Real Prices of Primary Commodities: Deflator Adjustment and the Prebisch-Singer Hypothesis," Working Paper, August 29, 2007 (updated September 14, 2007).

46. See Harvey et al., "The Prebisch-Singer Hypothesis," 367–377, at 376.

47. Marian Radetzki, *A Handbook of Primary Commodities in the Global Economy* (Cambridge: Cambridge University Press, 2008), 75–76; John O'Connor and David Orsmond, "The Recent Rise in Commodity Prices: A Long-run Perspective," *Reserve Bank of Australia Bulletin* (April 2007), 1–2.

48. Hubbert, for his part, argued in a major piece for *Scientific American* in 1955 that booming demand, even in the face of new discoveries, would deplete the world's reserves of a host of other exhaustible goods—lead, zinc, tin, gold, silver, and platinum—by the late 1980s, then copper by 2001. Copper, like oil, has been a favorite object of depletionist prediction-makers since the late nineteenth century. See Gorelick, *Oil Panic and the Global Crisis*.

49. BP, *BP Statistical Review of World Energy June 2012* (London, 2012), http://www.bp.com/statisticalreview.

50. Eyal Dvir and Kenneth S. Rogoff, "Three Epochs of Oil," NBER Working Paper No. 14927, National Bureau of Economic Research (April 2009).

51. Marian Radetzki, *Handbook of Primary Commodities in the Global Economy*, Reissue edition (New York: Cambridge University Press, 2010), 159-165.

52. BP, *BP Statistical Review of World Energy June 2012* (London: 2012), http://www.bp.com/statisticalreview. Figures include crude oil, shale oil, oil sands, and NGLs (the liquid content of natural gas where this is recovered separately).

53. See Robert Shiller, *Irrational Exuberance*, 1st and 2nd ed. (Princeton: Princeton University Press, 2000, 2005).

54. Robert Shiller, *Irrational Exuberance*, 2nd ed. paperback (New York: Crown Business, 2006), 106.

55. Ibid., 106.

56. Ibid., 107.

57. Ibid., 108.

58. Ibid.

59. Shiller defines a boom as a "major peak in the price-earnings ratio."

60. The primary sources cited in describing the excitement surrounding these bull markets are drawn from Shiller's *Irrational Exuberance*. They are included here to provide a sense of new era thinking, as he characterizes it, during these periods.

61. Ibid., 114.

62. Ibid., 126-129.

63. Ibid., 131.

64. A. J. Hazlitt, "Tremendous Oil Possibilities of Great Northwest Texas Region," *Oil Trade Journal* 9 (February 1918), 35.

65. Diana Davids Olien and Roger M. Olien, "Running Out of Oil: Discourse and Public Policy 1909-1929," *Business and Economic History* 22, no. 2 (Winter 1993), 36-66, at 38.

66. See James D. Hamilton, "Causes and Consequences of the Oil Shock of 2007-08," *Brookings Papers on Economic Activity* (Spring 2009), 23-24; James L. Smith, "World Oil: Market or Mayhem?" *Journal of Economic Perspectives* 23, no. 3 (Summer 2009), 145-164.

67. For an excellent study of how financial speculation affects oil prices, and how the market's evolution over the last decade has influenced price formation, see Bassam Fattouh, Lutz Kilian, and Lavan Mahadeva, "The Role of Speculation in Oil Markets: What Have We Learned So Far?" Working Paper, June 30, 2012.

68. George A. Akerlof and Robert J. Shiller, *Animal Spirits: How Human Psychology Drives the Economy and Why it Matters for Global Capitalism* (Princeton: Princeton University Press, 2010), 55.

69. Edward Morse and Michael Waldron, "Oil Dot-com," Lehman Brothers Energy Special Report, May 29, 2008.

70. Akerlof and Shiller, *Animal Spirits*.

71. Deffeyes, *Hubbert's Peak*, 1.

72. Dvir and Rogoff, "Three Epochs of Oil."

73. Ibid.

74. These annual price data come from the *BP Statistical Review of World Energy 2010* and *BP Statistical Review of World Energy 2011*, http://www.bp.com/statisticalreview. Its price series is derived from three different sets of annual average benchmark crude oil prices, which date back to 1861. The prices for the years 1861-1944 are based on U.S. average spot prices, while 1945-1983 is based on Arabian Light prices as posted at Ras Tanura, and 1984-2010 are the Brent dated prices.

75. U.S. Energy Information Administration, "U.S. Crude Proved Reserves (Million Barrels)," released August 2, 2012. Also see proved oil reserves history in BP, *BP Statistical Review of World Energy 2012* (London, June 2012), http://www.bp.com/statisticalreview.

76. Gorelick, *Oil Panic and the Global Crisis*, 142–143.

77. See Edward L. Morse et al., *Energy 2020: Independence Day*, Citi GPS: Global Perspectives and Solutions, February 2013, 53–55.

Chapter 02

1. U.S. Energy Information Administration, "U.S. Field Production of Crude Oil," Release date March 15, 2013. Accessible at http://www.eia.gov.

2. Benjamin Kline, *First along the River: A Brief History of the U.S. Environmental Movement*, 3rd ed. (Lanham, MD: Rowman & Littlefield Publishers, Inc., 2007), 51.

3. Harold F. Williamson, "Prophecies of Scarcity or Exhaustion of Natural Resources in the United States," *American Economic Review* 35, no. 2 (May 1945), 97–109, at 102.

4. Ibid.

5. Quoted in H. W. Brands, *T.R.: The Last Romantic* (New York: Basic Books, 1997), 624.

6. Roderick Frazier Nash, *American Environmentalism: Readings in Conservation History*, 3rd ed. (New York: McGraw-Hill, 1990), 84–89.

7. Theodore Roosevelt, "Seventh Annual Message to Congress," December 3, 1907, http://www.pbs.org/weta/thewest/resources/archives/eight/trconserv.htm.

8. Kirk Johnson, "From a Woodland Elegy, A Rhapsody in Green; Hunter Mountain Paintings Spurred Recovery," *New York Times*, June 7, 2001, http://www.nytimes.com/2001/06/07/nyregion/woodland-elegy-rhapsody-green-hunter-mountain-paintings-spurred-recovery.html?pagewanted=all&src=pm. Char Miller, *Gifford Pinchot and the Making of Modern Environmentalism* (Washington, DC: Island Press, 2001), 155.

9. U.S. Forest Service, "Gifford Pinchot (1865–1946)," Historical Information, http://www.fs.fed.us/gt/local-links/historical-info/gifford/gifford.shtml.

10. Diana Davids Olien and Roger M. Olien, "Running Out of Oil: Discourse and Public Policy 1909–1929," *Business and Economic History* 22, no. 2 (Winter 1993), 36–66, at 41.

11. Ibid., 41–42.

12. Also known as David T. Day, "The Petroleum Resources of the United States," in U.S. Geological Survey, *Papers on the Conservation of Mineral Resources*, Bulletin 394 (Washington, DC: Government Printing Office, 1909).

13. Ralph Arnold, "The Petroleum Resources of the United States," *Economic Geology* 10, no. 8 (December 1915), 695–712. Reprinted in *Annual Report of Board of Regents of the Smithsonian Institution Showing the Operations, Expenditures, and Condition of the Institution for the Year Ending June 30, 1916* (Washington, DC: Government Printing Office, 1917), 273.

14. *Report of the National Conservation Commission*, Volume I, Senate Document No. 676, 60th Congress, Second session (Washington, DC: Government Printing Office, 1909), 100.

15. David T. Day, "The Petroleum Resources of the United States," in Albert Shaw (ed.), *The American Review of Reviews* 39 (January-June 1909), 50.

16. Day, "The Petroleum Resources of the United States," in U.S. Geological Survey, 45.

17. *Report of the National Conservation Commission*, Volume I, 100.

18. Ibid.

19. Day, "The Petroleum Resources of the United States," in Shaw, 50.

20. Ibid.

21. The three publications broadcasting Day's findings—a U.S. Geological Survey bulletin, the American Review of Reviews, and the Report of the National Conservation Commission—were all published within months of each other. Day was the author of the first two of these publications; the third, in contrast, appears to be another author's summarization of his views (see *Report of the National Conservation Commission*, Volume I, 100). Yet the numbers they provide, and hence the conclusions they draw about the timing of the impending exhaustion of U.S. oil supplies, all vary somewhat. Rather than 1920 or 1935, for instance, the National Conservation Commission concludes more modestly that it

will occur "before the middle of the present century" (*Report of the National Conservation Commission*, Volume I, 100).

22. Day, "The Petroleum Resources of the United States," in U.S. Geological Survey, 47–50.

23. *Report of the National Conservation Commission*, Volume I, 1–8.

24. Olien and Olien, "Running Out of Oil," 43.

25. Ibid., 44.

26. G. O. Smith, "Our Mineral Resources," *Annals of the American Academy of Political and Social Science* (Philadelphia: American Academy of Political and Social Science, 1909), 195.

27. Ibid., 201.

28. "Gasoline to Jump Seven Cents a Gallon on First Day of Year," *Christian Science Monitor*, December 16, 1912.

29. "Gasoline Prices May Rise with Demand," *The Hartford Courant*, July 14, 1910.

30. "Oil Demand is Exceeding the Rate of Production," *Christian Science Monitor*, March 14, 1913.

31. U.S. Senate, *Report of the Senate Select Committee on Interstate Commerce*, Senate Report 46 Part 1, 49th Congress, First session (Washington, DC: Government Printing Office, 1886), 2–3.

32. Daniel Yergin, *The Prize: The Epic Quest for Oil, Money, and Power* (New York: Simon & Schuster, 1992), 54.

33. Gilbert H. Montague, *The Rise and Progress of the Standard Oil Company* (New York: Harper and Brothers, 1903), 130–132.

34. U.S. Federal Trade Commission, *Report of the Federal Trade Commission on the Pacific Coast Petroleum Industry: Part II, Prices and Competitive Conditions* (Washington, DC: Government Printing Office, 1922), 129.

35. Harold F. Williamson, Ralph L. Andreano, Arnold R. Daum, and Gilbert C. Klose, *The American Petroleum Industry: 1899–1959, the Age of Energy* (Evanston, IL: Northwestern University Press, 1963), 235.

36. A. G. Maguire, *Prices and Marketing Practices Covering the Distribution of Gasoline and Kerosene throughout the United States*, U.S. Fuel Administration, Oil Division (Washington: Government Printing Office, 1919), 7, 11.

37. "General Outlook for Oil," *Oil and Gas Journal*, October 10, 1912, 1; "Pinchot is for Conservation," *Oil and Gas Journal*, March 2, 1916, 2.

38. Olien and Olien, "Running Out of Oil."

39. "When the Wells Run Dry," *Oil and Gas Journal*, February 24, 1916, 2.

40. See Erik J. Dahl, "Naval Innovation: From Coal to Oil," *North American Shale Quarterly, E&P Magazine* (July 4, 2006), http://www.epmag.com/archives/digitalOilField/5911.htm.

41. "Letter to President Woodrow Wilson from Secretary of State William Jennings Bryan Regarding the Importance of U.S.-Occupied Oil Fields in Tampico, Mexico, Upon the Petroleum Needs of the Southwestern Section of the U.S., April 9, 1914., Record Group 130: Records of the White House Office, Records Relating to the Activities of the President and the White House Office, 1900–1935, Mexico, 1913–1916, National Archive." As cited in Roger Stern, "Oil Scarcity Ideology in US National Security Policy, 1909-1980," Working Paper on the Oil, Energy & the Middle East Program, Princeton University (2012), 2.

42. U.S. Energy Information Administration, "Table 4.2 Crude Oil and Natural Gas Cumulative Production and Proved Reserves, 1977–2010," Annual Energy Review (Release date September 27, 2012). http://www.eia.gov.

43. M. L. Requa, "An Article on the Exhaustion of the Petroleum Resources of the United States, Showing the Present and Future Supply and Demand, also the Production of the Principal Oil Fields of the United States," *Petroleum Resources of the United States*, U.S. Senate Document No. 363, 64th Congress, First session (Washington, DC: Government Printing Office, 1916).

44. Williamson et al., *The American Petroleum Industry*, 37.

45. Helmut Mejcher, *Imperial Quest for Oil: Iraq 1910–1928*, Middle East Centre, St. Antony's College (Oxford, UK: Ithaca Press, 1976), 37.

46. Eric D. K. Melby, *Oil and the International System: The Case of France, 1918–1969* (New York: Arno Press, 1981), 8–20; Henry Bérenger, *Le Pétrole et la France* (Paris: Flammarion, 1920), 41–55 (in French). Cited in Yergin, *The Prize*, 177.

47. Leonardo Maugeri, *The Age of Oil: The Mythology, History, and Future of the World's Most Controversial Resource* (Westport, CT: Praeger, 2006), 25.

48. Williamson et al., *The American Petroleum Industry*, 189–190.

49. "Model T Ford Production," Compiled by R. E. Houston, Ford Production Department, August 3, 1927, http://www.mtfca.com/encyclo/fdprod.htm; Douglas Brinkley, *Wheels for the World: Henry Ford, his Company, and a Century of Progress, 1903–2003* (New York: Penguin, 2003), 121–122, 272–273.

50. Williamson et al., *The American Petroleum Industry*, 195.

51. Arnold, "The Petroleum Resources of the United States," 274–287.

52. "Scarcity of Gas Oil Threatened," *Los Angeles Times*, December 27, 1914.

53. Mejcher, *Imperial Quest for Oil*, 37.

54. "Oil Shortage Seen for United States," *Christian Science Monitor*, March 15, 1918.

55. Chester G. Gilbert and Joseph E. Pogue, "Petroleum: A Resource Interpretation," Bulletin 102, Part 6 PL. 1, Smithsonian Institution, United States National Museum (Washington, DC: Government Printing Office, 1918).

56. Williamson et al., *The American Petroleum Industry*, 284–293.

57. "Careful Investigation Shows Gasoline Shortage Does Not Exist," *Los Angeles Times*, January 20, 1918.

58. "Gasoline Scarcity Rumor Questioned," *Christian Science Monitor*, September 18, 1918.

59. "Save Coal to Meet Industry's Greatest Peace Demand," *Power* 48, no. 22 (November 26, 1918), 74–75.

60. See, for instance, "War Hits Nearer while Oil Fuel Famine Impends," *The Hartford Courant*, May 22, 1918.

61. Maugeri, *The Age of Oil*, 25.

62. "Gasoline for Farmers," *Indiana Farmer's Guide*, March 9, 1918.

63. "Present Gasoline Supply Ample for All," *Motor Age*, July 18, 1918.

64. Edward G. Acheson, "Why Coal and Oil Conservation," *The Forum* (May 1919), 575–580.

65. Yergin, *The Prize*, 218.

66. "Gasoline Scarcity Rumor Questioned," *Christian Science Monitor*, September 18, 1918.

67. John A. DeNovo, "The Movement for an Aggressive American Oil Policy Abroad, 1918–1920," *American Historical Review* 61, no. 4 (July 1956), 854–876, 856–857.

68. David White, "The Petroleum Resources of the World," *Annals of the American Academy of Political and Social Science* 89 Prices (May 1920), 111–134, 132–133.

69. "Oil Burning in the Ships," *Oil & Gas Journal* 17, no. 29 (December 20, 1918), 2.

70. "Memorandum for the President of the United States from H. A. Garfield Concerning the Fuel Oil Situation, Josephus Daniels Papers, 518, Reel 36, Library of Congress." As cited in Roger Stern, "Oil Scarcity Ideology in US National Security Policy, 1909-1980," Working Paper of the Oil, Energy & the Middle East Program, Princeton University (2012), 5–6.

71. "Cadman to Fraser, December 2, 1920, 4247, Cadman papers." As cited in Yergin, *The Prize*, 195.

72. Olien and Olien, "Running Out of Oil," 49–50.

73. A. J. Hazlett, "Tremendous Oil Possibilities of Great Northwest Texas Oil Region," *The Oil Trade Journal* (February 1918), 35.

74. Yergin, *The Prize*, 209.

75. "Fuel Scarcity," *Wall Street Journal*, October 27, 1920.

76. Albert G. Robinson, "The Fuel of the Future," *Outlook*, July 21, 1920.

77. White, "The Petroleum Resources of the World," 111–134.

78. "Gasoline Famine Panic Exploded by U.S. Reports," *Chicago Daily Tribune*, June 27, 1920.

79. "Giebelhaus, *Sun*, p. 118." As cited in Yergin, *The Prize*, 222–223.
80. Quoted in Yergin, *The Prize*, 218.
81. U.S. Energy Information Administration, "U.S. Ending Stocks excluding SPR of Crude Oil," Release date March 15, 2013, http://www.eia.gov.
82. Williamson et al., *The American Petroleum Industry*, 303.
83. "Nation Faces Oil Famine," *Los Angeles Times*, September 23, 1923.
84. "Coolidge Appoints 4 Cabinet Members an Oil-Saving Board," *New York Times*, December 20, 1924.
85. Maugeri, *The Age of Oil*, 41–43.
86. Yergin, *The Prize*, 219.
87. Maugeri, *The Age of Oil*, 43.
88. Williamson et al., *The American Petroleum Industry*, 324.
89. Maugeri, *The Age of Oil*, 30–32.
90. American Petroleum Institute, *Petroleum Facts and Figures*, 374.
91. "Oil Famine Alarmists," *Los Angeles Times*, March 6, 1926.

Chapter 03

1. Harold F. Williamson, Ralph L. Andreano, Arnold R. Daum, and Gilbert C. Klose, *The American Petroleum Industry: 1899–1959, the Age of Energy* (Evanston, IL: Northwestern University Press, 1963), 796.
2. American Petroleum Institute, *Petroleum Facts and Figures: Centennial Edition* (New York: American Petroleum Institute, 1959), 374.
3. U.S. House of Representatives, *Hearings before a Subcommittee of the Committee on Interstate and Foreign Commerce*, House of Representatives, 77th Congress, Second session (Washington, DC: U.S. Government Printing Office, 1942), 4–6; The Advisory Commission to the Council of National Defense, *Defense* 2, no. 13 (April 1941); Official Weekly Bulletin of the Office of Emergency Management (March 1941), 4–5.
4. John W. Frey and H. Chandler Ide, *A History of the Petroleum Administration for War, 1941–1945* (Washington, DC: United States Government Printing Office, 1946), 172.
5. "Oil: Famine Closer," *Time* 38, no. 4 (July 28, 1941), 68.
6. "Epithets of the Week," *Life* 9, no. 10 (September, 2, 1940), 22.
7. Leonardo Maugeri, *The Age of Oil: The Mythology, History, and Future of the World's Most Controversial Resource* (Westport, CT: Praeger, 2006), 52.
8. "Ickes 'Gas' Shortage Scare Due to Anxiety Over Future," *Christian Science Monitor*, September 19, 1941.
9. Frey and Ide, *A History of the Petroleum Administration for War*, 119.
10. "Ickes Urges National D.S.T.; Power, Oil Shortage Foreseen," *Christian Science Monitor*, May 31, 1941.
11. Frey and Ide, *A History of the Petroleum Administration for War*, 119.
12. "Mayor Warns of Probable Fuel Oil Shortage," *Hartford Courant*, June 4, 1941.
13. "Oil Shortage has Arrived, U.S. Aid Warns," *Chicago Tribune*, August 21, 1941.
14. "Oil Shortages in the East Phony, Says Rep. Fish," *Chicago Daily Tribune*, August 26, 1941.
15. "A Shortage, but not of Gas," *Chicago Tribune*, August 25, 1941.
16. "The Gasoline Anti-Climax," *Hartford Courant*, October 24, 1941.
17. "Senate Committee Finds No Gasoline Shortage," *Los Angeles Times*, September 12, 1941.
18. "Davies Depicts Gasoline Crisis," *New York Times*, August 29, 1941.
19. Frey and Ide, *A History of the Petroleum Administration for War*, 119.
20. Ibid., 85–86.
21. "Mayor Warns of Probable Fuel Oil Shortage," *Hartford Courant*, June 4, 1941.
22. "Says Oil Men Seek to Frighten Buyers," *New York Times*, June 15, 1941.
23. "Moffett Hints U.S. Control of Oil Motivates Claims Shortage Exists," *Wall Street Journal*, October 4, 1941.

24. "Home Front: Oil or No Oil," *Time*, September 15, 1941; see also "A Shortage, but not of Gas," *Chicago Tribune*, August 25, 1941.

25. Frey and Ide, *A History of the Petroleum Administration for War*, 84–88.

26. Maugeri, *The Age of Oil*, 53.

27. Frey and Ide, *A History of the Petroleum Administration for War*, 185.

28. Williamson et al., *American Petroleum Industry*, 783–793.

29. Aaron Wildavsky and Ellen Tenenbaum, *The Politics of Mistrust: Estimating American Oil and Gas Resources* (Beverly Hills, CA: Sage Publications, 1981), 97–98.

30. Leonard M. Fanning, "A Case History of Oil-Shortage Scares," in Leonard M. Fanning (ed.), *Our Oil Resources*, 2nd ed. (New York: McGraw-Hill, 1950), 355.

31. Williamson et al., *American Petroleum Industry*, 770.

32. "U.S. as an Oil-Importing Nation is Forecast by Technologist," *New York Times*, July 15, 1943.

33. "Military Oil Needs will Increase; Civilians to Get Less—Ickes," *National Petroleum News*, February 24, 1943.

34. "William B. Heroy, 'The Supply of Crude Petroleum Within the United States,' July 29. 1943, pp. 4–9, 3417, DeGolyer papers." As cited in Daniel Yergin, *The Prize: The Epic Quest for Oil, Money, and Power* (New York: Simon & Schuster, 1992), 395.

35. "Out of Gas?" *Time* 42, no. 1 (July 5, 1943), 21.

36. Williamson et al., *American Petroleum Industry*, 779.

37. "Serious Petroleum Shortage Predicted in Near Future," *Science News Letter*, August 31, 1935. For the 1940 to 1943 time horizon specified, see the article on the same page, "Fears of Petroleum Shortage Held Greatly Exaggerated."

38. "Fears of Petroleum Shortage Held Greatly Exaggerated," *Science News Letter*, August 31, 1935.

39. "300-Year Oil Supply Believed in America," *New York Times*, September 26, 1943.

40. American Petroleum Institute, *Petroleum Facts and Figures*, 22.

41. "Hallanan Calls Crude Oil Price Rise Necessary to Avert Shortage," *Wall Street Journal*, May 10, 1943.

42. Frey and Ide, *A History of the Petroleum Administration for War*, 186; Wildavsky and Tenenbaum, *The Politics of Mistrust*, 100–101.

43. "Prentiss Brown (n.d.) quoted E. E. DeGolyer." As quoted in Wildavsky and Tenenbaum, *The Politics of Mistrust*, 100–101.

44. "Plymouth Oil President Warns of Crude Shortages," *Wall Street Journal*, March 26, 1943.

45. "Prentiss Brown (n.d.) wrote to Harold Ickes." Cited in Wildavsky and Tenenbaum, *The Politics of Mistrust*, 101.

46. Investigation of the National Defense Program, *Additional Report: Report of Subcommittee Concerning Investigations Overseas*, Report 10, Part 15, Senate, 78th Congress, Second session (Washington, DC: Government Printing Office, 1944), 527, 550. Reprinted in Investigation of the National Defense Program, *Additional Report of the Special Committee Investigating the National Defense Program*, Report No. 10, Part 16, Senate, 78th Congress, Second session (Washington, DC: Government Printing Office, 1944).

47. Maugeri, *The Age of Oil*, 54.

48. E. L. DeGolyer, "Preliminary Report of the Technical Oil Mission to the Middle East: 1 Feb. 1944, Roosevelt Papers, OF 4226-D." As cited in David S. Painter, *Oil and the American Century: The Political Economy of US Foreign Oil Policy, 1941–1954* (Baltimore: Johns Hopkins University Press, 1986), 52; John H. Murrell, "Middle East Oil," *Tulsa Geological Society Digest* 14 (1945–1946), 48–49.

49. Yergin, *The Prize*, 398.

50. Investigation of the National Defense Program, *Additional Report*, 506–508, 515–518, 579. Reprinted in Investigation of the National Defense Program, *Additional Report of the Special Committee Investigating the National Defense Program*, Report No. 10, Part 16, Senate, 78th Congress, Second session (Washington, DC: Government Printing Office, 1944).

51. "Arabian Oil Plan Defended by Knox," *New York Times*, March 22, 1944.

52. "Roosevelt Links Pipeline to Needs," *New York Times*, March 4, 1944.

53. As quoted in Fanning, "A Case History of Oil-Shortage Scares," 360.

54. Ralph Zook, *The Proposed Arabian Pipeline: A Threat to Our National Security* (Tulsa, OK: Independent Petroleum Association of America, 1944).

55. Frey and Ide, *A History of the Petroleum Administration for War*, 444–445.

56. As described in Yergin, *The Prize*, 406.

57. "Self-Sufficiency Stressed in Senate Oil Policy Report," *National Petroleum News*, February 5, 1947, 12.

58. "Forrestal to Secretary of State, December 11, 1944, 890F.6363/12-1144." As cited in Yergin, *The Prize*, 406–407.

59. "Collado to Clayton, March 27, 1945, 890F.6363/3-2745, RG 59, NA. Walter Millis, ed., *The Forrestal diaries* (New York: Viking, 1951) p. 81." As cited in Yergin, *The Prize*, 407.

60. United States Tariff Commission, *Petroleum: Prepared in Response to Requests from the Committee on Finance of the United States and the Committee on Ways and Means of the House of Representatives*, War Changes in Industry Series no. 17 (Washington, DC: United States Government Printing Office, 1946), 7.

61. Yergin, *The Prize*, 402.

62. Quote from Fanning, "A Case History of Oil-Shortage Scares," 365–366.

63. "Pointing to our Need for Foreign Oil Supply Places Industry at Fork of Road to Control," *National Petroleum News*, August 21, 1946, 23–24.

64. "Krug Tells Compact Substantial Oil Imports will be Needed," *Oil and Gas Journal*, August 17, 1946, 67–68.

65. American Petroleum Institute, *Petroleum Facts and Figures*, 209–211.

66. Yergin, *The Prize*, 409.

67. Quoted in Fanning, "A Case History of Oil-Shortage Scares," 379–380.

68. Yergin, *The Prize*, 410.

69. "Standard Oil (Ind.) Rationing Gas in Midwest," *Chicago Tribune*, June 25, 1947.

70. "Gas, Fuel Oil Pinch Looms, Krug Says," *New York Times*, June 18, 1947.

71. "Oil: Summer Shortage," *Time* 50, no. 1 (July 7, 1947), 85.

72. "Supply Picture Unchanged Despite 'Scare' Headlines," *National Petroleum News*, June 25, 1947.

73. "Nation Warned It Faces Oil and Gas Shortage," *Chicago Daily Tribune*, June 18, 1947.

74. "Executive Sees No Danger of Oil Shortage," *Christian Science Monitor*, July 17, 1947.

75. Fanning, "A Case History of Oil-Shortage Scares," 370.

76. Yergin, *The Prize*, 409.

77. "Oil Companies Discount Reports of Gas Shortage," *Christian Science Monitor*, October 14, 1947.

78. "U.S. and Europe Face Oil Shortage," *New York Times*, November 16, 1947.

79. "Oil Companies Discount Reports of Gas Shortage," *Christian Science Monitor*, October 14, 1947.

80. "Krug Warns Oil Shortage Here to Stay," *Christian Science Monitor*, February 16, 1948.

81. "President Orders Federal Heat Cut and Autos Slowed," *New York Times*, January 18, 1948.

82. "Probe Sought in Bay State Oil Shortage," *Christian Science Monitor*, January 13, 1948.

83. "Oil: Cold Comfort," *Time*, February 2, 1948; "Oil: Petroleum Economy," *Time*, February 16, 1948.

84. "Oil Curbs Seen Prolonging Shortage," *Christian Science Monitor*, January 29, 1948.

85. "Oil Men Face Dilemma in Meeting Shortages," *Christian Science Monitor*, February 25, 1948.

86. "No Crisis in Oil Seen by Holman Who Says Supply to Equal Demand," *Oil and Gas Journal*, June 10, 1948. The final portion of the quote does not appear in the article but is given in Fanning, "A Case History of Oil-Shortage Scares," 391–392.

87. "Oil: Cold Comfort," *Time*, February 2, 1948.

88. "Shortage Expected to Last Five Years," *Christian Science Monitor*, February 20, 1948.

89. Ibid.

90. Samuel A. Tower, "Oil Running Short, House Body Warns," *New York Times*, May 7, 1948.

91. "Large Oil Supply Seen," *New York Times*, May 7, 1948.

92. Hearing before the Committee on Foreign Relations on the Nomination of Dean G. Acheson to be Secretary of State, United States Senate, 81st Congress, First Session (Washington, DC: United States Government Printing Office, 1949), 466.

93. Hearings before the Committee on Foreign Affairs on H.R. 2362 A Bill to Amend the Economic Cooperation Act of 1948, Part 1, United States House of Representatives, 81st Congress, First Session (Washington, DC: United States Government Printing Office, 1949), 510.

94. Williamson et al., *American Petroleum Industry*, 811.

95. Yergin, *The Prize*, 430.

96. American Petroleum Institute, *Petroleum Facts and Figures*, 62–63.

97. "Shortages of Oil Disappear—Stocks on Hand Accumulate," *Christian Science Monitor*, January 3, 1949.

98. "Ohio Oil President Sees Petroleum Shortages a Thing of the Past," *Wall Street Journal*, May 27, 1949.

99. Williamson et al., *American Petroleum Industry*, 810.

100. "Oil: Quick Change," *Time* 56, no. 4 (July 24, 1950), 76.

Chapter 04

1. BP, *BP Statistical Review of World Energy June 2012* (London, BP, June 2012).

2. James E. Akin, "The Oil Crisis: This Time the Wolf Is Here," *Foreign Affairs* 51, no. 3 (April 1973), 462–490, at 462–463.

3. BP, *BP Statistical Review of World Energy June 2012*.

4. Leonardo Maugeri, *The Age of Oil: The Mythology, History, and Future of the World's Most Controversial Resource* (Westport, CT: Praeger, 2006), 84.

5. Ian Skeet, *OPEC: Twenty-Five Years of Prices and Politics* (Cambridge: Cambridge University Press, 1988).

6. Maugeri, *The Age of Oil*, 85.

7. Ibid., 107.

8. Walter J. Levy, "Oil Power," *Foreign Affairs*, July 1971.

9. Francisco Parra, *Oil Politics: A Modern History of Petroleum* (New York: I. B. Taurus and Co., 2010), 148–149, 161.

10. "Areas of Growth," *Petroleum Press Service* (July 1969), 242–244.

11. Parra, *Oil Politics*, 117.

12. Daniel Yergin, *The Prize: The Epic Quest for Oil, Money, and Power* (New York: Simon & Schuster, 1992), 590.

13. Morris A. Adelman, *The Genie out of the Bottle: World Oil since 1970* (Cambridge, MA: MIT Press, 1995), 85.

14. Department of State, "The International Oil Industry through 1980," December 1971, in Muslim Students Following the Line of the Imam, *Documents from the U.S. Espionage Den*, vol. 57 (Tehran: Center for the Publication of the U.S. Espionage Den's Documents, 1986).

15. BP, *BP Statistical Review of World Energy June 2012*.

16. "U.S. Official Warns of Imports Danger," *Oil and Gas Journal*, May 15, 1972, 50.

17. Clyde H. Farnsworth, "Trading Bloc Told Oil Dearth Looms," *New York Times*, May 26, 1972.

18. Rachel Carson, *Silent Spring* (Boston: Houghton Mifflin, 1962).

19. "Earth Day: The History of a Movement," Earth Day Network, http://www.earthday.org/earth-day-history-movement.

20. Donella H. Meadows, et al., *Limits to Growth: A Report for the Club of Rome's Project on the Predicament of Mankind*, 2nd ed. (New York: Universe Publishing, 1974).

21. BP, *BP Statistical Review of World Energy 2011* (London BP, 2011), http://www.bp.com/statisticalreview.
22. Vaclav Smil, *Energy at the Crossroads* (Cambridge, MA: MIT Press, 2003).
23. Yergin, *The Prize*, 569.
24. *Petroleum Intelligence Weekly*, September 10, 1973, 12.
25. *Petroleum Intelligence Weekly*, August 13, 1973, 3–4.
26. Quoted text from Maugeri, *The Age of Oil*, 108.
27. Ibid., 107; Yergin, *The Prize*, 591.
28. Parra, *Oil Politics*, 179.
29. Parra gives slightly differing figure, claiming the price increase was from $3.01 to $5.11 per barrel. Parra, *Oil Politics*, 178–179.
30. Ibid.; Maugeri, *The Age of Oil*, 114; Yergin, *The Prize*, 615.
31. Parra, *Oil Politics*, 183.
32. Adelman, *Genie out of the Bottle*, 110.
33. Yergin, *The Prize*, 615.
34. "U.S. Urges Allied Unity on Oil Crisis," *The Spokesman-Review*, December 13, 1973.
35. Adelman, *Genie out of the Bottle*, 110.
36. Bernard Weintraub, "Iran Keeps Oil Flowing despite Reported Pressure from Arabs," *New York Times*, December 18, 1973.
37. See "How Scarce is Oil," *Economist*, December 15, 1973.
38. Maugeri, *The Age of Oil*, 113. Yergin puts these figures slightly higher. He calculates that the net loss of supply constituted 9 percent of the total barrels per day relative to the quantity supplied in September 1973, or 14 percent of the total global trade volume.
39. See U.S. Energy Information Administration, "U.S. Ending Stocks of Total Gasoline (Thousand Barrels)" and "U.S. Ending Stocks of Distillate Fuel Oil (Thousand Barrels)," Release date: May 30, 2013.
40. Adelman, *Genie out of the Bottle*, 112.
41. Henry Kissinger, *Years of Upheaval: The Second Volume of His Classic Memoirs* (New York: Simon & Shuster Paperbacks, 2011), 873.
42. Quoted in Jerry Taylor and Peter Van Doren, "An Oil Embargo Won't Work," *Wall Street Journal*, April 10, 2002.
43. Richard Nixon, "Address to the Nation about Policies to Deal with the Energy Shortages," November 7, 1973.
44. "Yankelovich to Haig, December 6, 1973, with memorandum." As quoted in Yergin, *The Prize*, 618.
45. "Shah of Iran Explains Latest Oil Price Increase," Supplement, *Middle East Economic Survey* 17, no. 10 (December 28, 1973), 3–4.
46. Steve Isser, *The Economics and Politics of the United States Oil Industry, 1920–1990: Profits, Populism, and Petroleum* (New York: Garland Publishing, Inc., 1996), 251.
47. "The Monthly Oil & Energy Trends (February 1979)," as cited in Parra, *Oil Politics*, 218.
48. Yergin, *The Prize*, 671.
49. United States Central Intelligence Agency, *The International Energy Situation: Outlook to 1985* (Washington, DC: Government Printing Office, 1977).
50. "Schlesinger Warns West U.S. Energy Plan Is Vital; Program to Ease Reliance on Imported Oil Adopted by Ministers of I.E.A.," *New York Times*, October 6, 1977.
51. Rockefeller Foundation, *Working Paper on International Energy Supply: An Industrial World Perspective* (New York: Rockefeller Foundation, 1978).
52. Parra, *Oil Politics*, 253.
53. Jimmy Carter, "The President's Proposed Energy Policy," April 18, 1977, http://www.pbs.org/wgbh/americanexperience/features/primary-resources/carter-energy.
54. U.S. Energy Information Administration, "Real Prices Viewer," http://www.eia.gov/forecasts/steo/realprices/.
55. See Parra, *Oil Politics*, 222–223; Yergin, *The Prize*, 685.
56. Parra, *Oil Politics*, 228.

57. Yergin, *The Prize*, 685–686.
58. Parra, *Oil Politics*, 218.
59. Yergin, *The Prize*, 686–687.
60. Robert Kagan, *The World America Made* (New York: Vintage Books, 2013), 118.
61. James R. Schlesinger, "Energy Risks and Energy Futures: Some Farewell Observations," Supplement, *Petroleum Intelligence Weekly*, August 27, 1979.
62. "Energy Risks and Energy Futures," *Wall Street Journal*, August 23, 1979.
63. British Petroleum, *Oil Crisis... Again, A Brief by the Policy Review Unit* (London: The British Petroleum Company Limited, September 1979). Cited in Leonardo Maugeri, *The Age of Oil: The Mythology, History, and Future of the World's Most Controversial Resource* (Westport, CT: Praeger, 2006), 130.
64. Exxon Corporation, 1979 Annual Report (1979), 2–3.
65. United States Central Intelligence Agency, *The World Oil Market in the Years Ahead* (Washington, DC: Government Printing Office, 1979).
66. "The CIA Reassess the Geopolitics of Oil," Special Supplement, *Petroleum Intelligence Weekly*, May 19, 1980.
67. "US View of OPEC–West Relations," *Petroleum Economist* (December 1979), 506.
68. Quoted in Parra, *Oil Politics*, 223.
69. U.S. Senate Committee on Energy and National Resources, "Part I—The Gathering Energy Crisis," *The Geopolitics of Oil*, Staff Report, 96th Congress, Second session (Washington, DC: Government Printing Office, December 1980).
70. Parra, *Oil Politics*, 216–217.
71. Maugeri, *The Age of Oil*, 134–136.
72. Stephen Haber, Noel Maurer, and Armando Razo, "When the Law Does Not Matter: The Rise and Decline of the Mexican Oil Industry," *Journal of Economic History* 63, no. 1 (March 2003), 1–31, at 1.
73. K. W. Glennie, *Introduction to the Petroleum Geology of the North Sea* (Oxford: Blackwell Scientific Publications, 1984).
74. Yergin, *The Prize*, 748.
75. Sheikh Ahmed Zaki Yamani, "Debate at the Oxford Energy Seminar, 13 September 1985," in Robert Mabro (ed.), *OPEC and the World Market: The Genesis of the 1986 Price Crisis* (Oxford: Oxford University Press—Oxford Institute for Energy Studies, 1986), 165–168.
76. Parra, *Oil Politics*, 287; Maugeri, *The Age of Oil*, 139.
77. Isser, *Economics and Politics of the United States Oil Industry*, 170.
78. Maugeri, *The Age of Oil*, 139, 296.

Chapter 05

1. "Oil Shocked," *Economist*, March 26, 1998.
2. U.S. Energy Information Administration, "International Energy Statistics," http://www.eia.gov.
3. For a sense of the market commentary during the 1997–98 price collapse, see Agis Salpukas, "Oil's Numbers Game," *New York Times*, December 1, 1997; "OPEC Nations Seen Lifting Ceiling on Oil Production," *New York Times*, November 27, 1997; Youssef M. Ibrahim, "Falling Oil Prices Pinch Several Producing Nations," *New York Times*, June 23, 1998. Also see the "special feature" section of the IEA's Monthly Oil Market Reports.
4. See the Trade-Weighted U.S. Dollar Index, published by the Board of Governors of the Federal Reserve System, http://research.stlouisfed.org/fred2/categories/105.
5. Keith Bradsher, "Vehicles to Vie for King of Hill," *New York Times* News Service, *Sarasota Herald Tribune*, June 17, 1997.
6. "Charge of the Large," News Service Detroit, *New York Times*, June 26, 1997.
7. "Merger Won't Fix Oil-Glut Fallout," *The Florida Times-Union*, December 2, 1998.
8. Daniel Yergin recounts these mergers and acquisitions in detail in *The Quest: Energy, Security, and the Remaking of the Modern World* (New York: Penguin Press, 2011), 87–105.
9. David E. Sanger, "Singing the Cartel Blues," *New York Times*, March 29, 1998.

10. Agis Salpukas, "Challenges Inside and Out Confront OPEC," *New York Times*, June 23, 1998.

11. Sanger, "Singing the Cartel Blues."

12. Daniel Yergin and Joseph Stanislaw, "How OPEC Lost Control of Oil," *Time* 151, no. 13 (April 6, 1998), 58.

13. The first public record of the stand-alone term "peak oil" in the popular media to describe the global point of maximum oil output appears in May 2002, following a conference held in Uppsala University in Sweden. Colin Campbell helped organize the conference. See, for instance, "Global Reserves of Crude Oil Could Peak by Year 2010, Say Int'l Experts," Associated Press, May 26, 2002.

14. Colin J. Campbell and Jean H. Laherrère, "The End of Cheap Oil," *Scientific American* (March 1998), 78–83.

15. John D. Edwards, "Crude Oil and Alternate Energy Production Forecasts for the Twenty-First Century: The End of the Hydrocarbon Era," *AAPG* Bulletin 81, no. 8 (August 1997), 1292–1305.

16. For a partial bibliography of Hubbert-style pieces on peak oil over the last few decades, see Chris Kuykendall, "M. King Hubbert and His Successors: A Half-Bibliography through 2005," March 2006, http://www.mkinghubbert.com/files/HubbertBibliography.20060308.pdf.

17. L. F. Ivanhoe, "Get Ready for Another Oil Shock!" *The Futurist* 31, no. 1 (January–February 1997), 20–23

18. Colin J. Campbell, *The Golden Century of Oil 1950–2050: The Depletion of a Resource* (New York: Springer, 1991), ix–x, 51–52.

19. C. J. Campbell, "The Twenty First Century: The World's Endowment of Conventional Oil and its Depletion," January 1996, http://www.oilcrisis.com/campbell/camfull.htm.

20. Taken from "On Peak Oil," a short introduction to the idea of peak oil written by Campbell. The piece is accessible at http://www.peakoil.net/about-peak-oil. In *Oil: Money, Politics, and Power in the 21st Century*, Tom Bowers says the piece dates from June 1996. See Tom Bowers, *Oil: Money, Politics, and Power in the 21st Century* (New York: Grand Central Publishing, 2010), 271.

21. Only later would he develop a more formal symmetric logistic distribution curve to fit the data. See Steven M. Gorelick, *Oil Panic and the Global Crisis: Predictions and Myths* (Chichester, U.K.: Wiley-Blackwell, 2010).

22. M. K. Hubbert, "Nuclear Energy and the Fossil Fuels," *Drilling and Production Practice*, American Petroleum Institute (1956), 7–25.

23. Rob Scherer, "High Oil Prices Siphon Profits from US Firms," *Christian Science Monitor*, September 21, 2000.

24. Amy Myers Jaffe and Robert A. Manning, "The Shocks of a World of Cheap Oil," *Foreign Affairs* 79, no. 1 (January/February 2000).

25. International Energy Agency, *Oil Market Report*, June 11, 2000, 4.

26. Brian Knowlton, "Gore Urges Use of Oil Reserves to Ease Prices," *New York Times*, September 22, 2000.

27. International Energy Agency, *Oil Market Report*, December 11, 2000, 3.

28. USGS World Assessment Team, "U.S. Geological Survey World Petroleum Assessment 2000—Description and Results," U.S. Geological Survey Digital Data Series—DDS-60, U.S. Geological Survey, U.S. Department of the Interior, http://pubs.usgs.gov/dds/dds-060/.

29. U.S. Energy Information Administration, "Long Term World Oil Supply: A Resource Base/Production Path Analysis," Presentation by Jay Hakes to the American Association of Petroleum Geologists on April 18, 2000, New Orleans, Louisiana. A version of this presentation was given at the April 18, 2000, meeting described. The presentation is accessible at http://www.eia.gov/pub/oil_gas/petroleum/presentations/2000/long_term_supply/sld001.htm.

30. International Energy Agency, *World Energy Outlook 1998 Edition* (Paris: IEA Publications, 1998), 3, 44–46, 91–113, 120–121.

31. International Energy Agency, *World Energy Outlook 2000* (Paris: IEA Publications, 2000), 22, 74, 76.

32. U.S. Energy Information Administration, *Short-Term Energy Outlook*, Release Date June 11, 2013. See U.S. Energy Information Administration, "Real Prices Viewer," http://www.eia.gov/forecasts/steo/realprices/.

33. "A National Report on America's Energy Crisis," Remarks by U.S. Secretary of Energy Spencer Abraham at the U.S. Chamber of Commerce, National Energy Summit, March 19, 2001.

34. Alison Mitchell, "Senator Promises Investigation into Gasoline Price Rise," *New York Times*, May 30, 2001.

35. Pew Research Center for the People & the Press, "From News Interest To Lifestyles, Energy Takes Hold," May 24, 2001.

36. For an excellent discussion of the effects of 9/11 on the oil market and the United States, see Yergin, *The Quest*, 125–129.

37. Kenneth S. Deffeyes, *Hubbert's Peak: The Impending World Oil Shortage* (Princeton, NJ: Princeton University Press, 2001), 6.

38. "Oil Depletion: Sunset for the Oil Business?" *Economist*, November 1, 2001.

39. William McCall, "Physicist Warns Global Oil Production Peak could Bring Economic Disaster," Associated Press, August 13, 2001.

40. The Arlington Institute, "The Oil Depletion Analysis Centre," http://www.arlingtoninstitute.org/oil-depletion-analysis-centre.

41. See chapter 8 in Charles A. S. Hall and Carlos A. Ramirez-Pascualli, *The First Half of the Age of Oil: An Exploration of the Work of Colin Campbell and Jean Laherrère*, Springer Briefs in Energy, Energy Analysis (New York: Springer, 2013).

42. Goldman Sachs, "Underinvestment in Commodities Means Markets will be Tighter, Sooner," CEO Confidential, Issue 2002/05 (April 2002).

43. Bowers, *Oil: Money, Politics, and Power*, 266–282.

44. See International Energy Agency, *Oil Market Report*, September 11, 2002, 3; *Oil Market Report*, December 11, 2002, 6; *Oil Market Report*, January 17, 2003, 4; *Oil Market Report*, March 15, 2002, 25.

45. Yergin, *The Quest*, 141–158.

46. International Energy Agency, *Oil Market Report*, March 12, 2003, 3.

47. For monthly oil supply figures, see U.S. Energy Information Administration, "International Energy Statistics," http://www.eia.gov.

48. IMF, "Gross Domestic Product, Constant Prices," in International Monetary Fund, World Economic Outlook Database, April 2013, http://www.imf.org/external/pubs/ft/weo/2013/01/weodata/index.aspx.

49. See BP, *BP Statistical Review of World Energy June 2012* (London, BP: June 2012), http://www.bp.com/statisticalreview.

50. IMF, "Gross Domestic Product, Constant Prices."

51. Yergin, *The Quest*, 216–217.

52. See BP, *BP Statistical Review of World Energy June 2012*.

53. Alan Heap, *China—The Engine of a Commodities Super Cycle*, Smith Barney, Citigroup Global Markets, March 31, 2005.

54. Yergin, *The Quest*, 163.

55. OPEC in this case refers to the so-called "OPEC 12." Production capacity data are Bloomberg estimates as of June 2013.

56. International Energy Agency, "OPEC on a Roll," *Oil Monthly Report: 10 October 2003* (October 10, 2003), 3.

57. Michael C. Lynch, "Crying Wolf: Warnings about Oil Supply," March 1998, http://sepwww.stanford.edu/sep/jon/world-oil.dir/lynch/worldoil.html.

58. Michael C. Lynch, "The New Pessimism about Petroleum Resources: Debunking the Hubbert Model (and Hubbert Modelers)," *Minerals & Energy—Raw Materials Report* 18, no. 1 (2003).

59. Amy Myers Jaffe, "Not So Cheap," *Foreign Affairs*, May 26, 2004, http://www.foreignaffairs.com/articles/64222/amy-myers-jaffe/not-so-cheap.

60. Jeff Gerth, "Forecast of Rising Oil Demand Challenges Tired Saudi Fields," *New York Times*, February 24, 2004.

61. Matthew R. Simmons, *Twilight in the Desert: The Coming Saudi Oil Shock and the World Economy* (Hoboken, NJ: Wiley, 2005), xvii.

62. Michael T. Klare, "The Saudi Oil Bombshell," Asia Times Online, June 29, 2005, http://www.atimes.com/atimes/Middle_East/GF29Ak01.html.

63. John Vidal, "Analyst Fears Global Oil Crisis in Three Years," *Guardian*, April 26, 2005.

64. Robert L. Hirsch, Roger Bezdek, and Robert Wendling, *Peaking of World Oil Production: Impacts, Mitigation, & Risk Management*, Report prepared by Science Applications International Corporation for the National Energy Technology Laboratory, U.S. Department of Energy (February 2005).

65. Padraic Cassidy, "Goldman Sees Oil Price 'Super Spike,'" MarketWatch, *Wall Street Journal*, March 31, 2005, http://www.marketwatch.com/story/goldman-sees-oil-super-spike-others-are-skeptical.

66. Shira Ovide, "A Career 'Super Spike' for Goldman's Infamous Oil Analyst," Deal Journal, *Wall Street Journal*, February 1, 2011, http://blogs.wsj.com/deals/2011/02/01/a-career-super-spike-for-goldmans-infamous-oil-analyst/.

67. Cassidy, "Goldman Sees Oil Price 'Super Spike.'"

68. Mark Shenk, "Oil Jumps to Record $58.60 as Demand May Outpace Production," *Bloomberg*, June 17, 2005.

69. "Oil 'Peak' Not Seen Coming Any Time Soon," Associated Press, June 21, 2005.

70. Joseph Nocera, "On Oil Supply, Opinions Aren't Scarce," *New York Times*, September 10, 2005.

71. Jad Mouawad, "With Oil Prices off their Peak, are Supplies Assured?," *New York Times*, December 5, 2005.

72. Michael J. Deslauriers, "Famed Oil Tycoon Sounds off on Peak Oil," *Resource Investor*, June 23, 2005, http://www.resourceinvestor.com/2005/06/23/famed-oil-tycoon-sounds-off-on-peak-oil.

73. Nocera, "On Oil Supply, Opinions Aren't Scarce."

74. John Mintz, "Outcome Grim at Oil War Game," *Washington Post*, June 24, 2005.

75. Katie Benner, "Lawmakers: Will We Run Out of Oil?," CNNMoney.com, December 7, 2005, http://money.cnn.com/2005/12/07/markets/peak_oil/.

76. Joe Carroll, "Global Oil Output Won't Peak until 2030, Yergin Says (Update1)," *Bloomberg*, November 14, 2006.

77. SF Informatics, "Technology Won't Solve America's Oil Addiction, Experts Say," Press Release, PRWeb, February 2, 2006, http://www.prweb.com/releases/200614/2/prweb341058.htm.

78. Bill Steigerwald, "Pickens: We Can't Become Energy Independent," *Pittsburgh Tribune-Review*, February 12, 2006.

79. Michael Hirsch, "The Energy Wars," *Newsweek*, May 3, 2006.

80. Stephen Voss, "Saudi, U.S. Officials Confident Technology will Boost Reserves," *Bloomberg*, September 13, 2006.

81. Kristen Hays, "Peak Oil Debate Crackles Anew," *Houston Chronicle*, November 15, 2006.

82. Joe Carroll, "Global Oil Output Won't Peak until 2030, Yergin Says (Update1)," *Bloomberg*, November 14, 2006.

83. International Energy Agency, *Oil Market Report*, October 11, 2007, 4.

84. William Patalon III, "Goldman Sachs Follows Money Morning Prediction That Oil Prices Could Approach $200 a Barrel," Money Morning, March 17, 2008, http://money-morning.com/2008/03/17/goldman-sachs-follows-money-morning-prediction-that-oil-prices-could-approach-200-a-barrel/.

85. Dylan Bowman, "Crude to hit $175, says Goldman Sachs," arabianbusiness.com, March 15, 2008, http://www.arabianbusiness.com/crude-hit-175-says-goldman-sachs-52,406.html.

86. Lawrence C. Strauss, "What Mr. Crude Oil Sees Ahead, " *Barron's*, June 9, 2008, http://online.barrons.com/article/SB121279317214553377.html#articleTabs_article%3D1.

87. Jad Mouawad, "Oil Price Rise Fails to Open Tap," *New York Times*, April 29, 2008.

88. Edward Morse and Michael Waldron, "Oil Dot-com," Lehman Brothers Energy Special Report, May 29, 2008.

89. Steve Hargreaves, "Why $120 Oil is Good," CNNMoney.com, May 8 2008, http://money.cnn.com/2008/05/07/news/economy/120_oil/.

90. Jad Mouawad, "The Big Thirst," New York Times, April 20, 2008.

91. Looking back at the epic round trip prices took in 2007 and 2008, some experts saw the hallmarks of a classic speculative bubble, though others were not so sure. For two excellent discussions of the question drawing on the econometric evidence, see James L. Smith, "World Oil: Market or Mayhem?" Journal of Economic Perspectives 23, no. 3 (Summer 2009), 145–164; and James D. Hamilton, "Cause and Consequences of the Oil Shock of 2007–08," Brookings Papers on Economic Activity (Spring 2009), 215–261.

92. Deborah Zabarenko, "Oil Running Out as Prime Energy Source: World Poll," Reuters, April 20, 2008.

93. See Google Zeitgeist 2008, "Top of Mind," http://www.google.com/intl/en/press/zeitgeist2008/mind.html.

94. "Paulson: 'No Quick Fix' to Oil Prices," USAToday, June 1, 2008.

95. Jad Mouawad, "Big Oil Projects Put in Jeopardy by Fall in Prices," New York Times, December 15, 2008.

96. Barclays Research Estimates. See Blake Clayton, "Drilling into the American Energy Boom, in Four Charts," Energy, Security, and Climate, CFR.org, January 8, 2013, http://blogs.cfr.org/levi/2013/01/08/drilling-into-the-american-energy-boom-in-4-charts/.

97. Brian O'Keefe, "Exxon's Big Bet on Shale Gas," CNNMoney.com, April 16, 2012, http://tech.fortune.cnn.com/2012/04/16/exxon-shale-gas-fracking/.

98. Clifford Krauss, "Shale Boom in Texas Could Increase U.S. Oil Output," New York Times, May 27, 2011.

99. Sheila McNulty, "Oil Back in Favor with US Drillers after Years of Targeting Gas," Energy Source, Financial Times, July 14, 2010, http://blogs.ft.com/energy-source/2010/07/14/oil-back-in-favor-with-us-drillers-after-years-of-targeting-gas/#axzz2ZFR6ZkAf.

100. U.S. Energy Information Administration, "EIA Tracks U.S. Tight Oil Production as Volumes Soar," This Week in Petroleum, Release Date March 14, 2012.

101. Clifford Krauss, "There will be Fuel," New York Times, November 16, 2010.

102. Amy Myers Jaffe, "The Americas, Not the Middle East, will be the World Capital of Energy," Foreign Policy, no. 188 (September/October 2011), 1–3.

103. Daniel Yergin, "There will be Oil," Wall Street Journal, September 27, 2011.

104. Amanda Scott, "Obama Talks Energy at the State of the Union 2013," Energy.Gov. http://energy.gov/articles/president-obama-talks-energy-state-union-2013.

105. See the IEA's 2013 Medium Term Oil Market Report, as well as the supplementary materials (including Maria van der Hoeven's remarks) referenced in the following online IEA press release: International Energy Agency, "Supply Shock from North American Oil Rippling through Global Markets," Press Release, May 14, 2013, http://www.iea.org/newsroomandevents/pressreleases/2013/may/name,38,080,en.html.

106. Colin Sullivan, "Has 'Peak Oil' Gone the Way of the Flat Earth Society," EnergyWire, March 22, 2013.

107. Charles C. Mann, "What If We Never Run Out of Oil?," The Atlantic, April 24, 2013.

Chapter 06

1. See Niels Jensen, "The Case for Human Ingenuity," Credit Writedowns, May 3, 2011, http://www.creditwritedowns.com/.

2. Alex Trembath, Jesse Jenkins, Ted Nordhaus, and Michael Shellenberger, Where the Shale Gas Revolution Came From: Government's Role in the Development of Hydraulic Fracturing in Shale, Breakthrough Institute, May 2012, http://thebreakthrough.org/.

3. Matthew R. Simmons, "Revisiting the Limits to Growth: Could the Club of Rome Have Been Correct, After All?" An Energy White Paper, Simmons & Company International, October 2000, 4–5. Available on Simmons & Company website, as of August 2014: http://www.simmonsco-intl.com/About-Us/Our-Founder/Legacy-Presentations/.

4. James D. Hamilton, "Understanding Crude Oil Prices," Working paper, May 22, 2008 (Revised December 6, 2008), 2–4, http://dss.ucsd.edu/~jhamilto/understand_oil.pdf.
5. Carmen M. Reinhart and Kenneth S. Rogoff, *This Time is Different: Eight Centuries of Financial Folly* (Princeton, NJ: Princeton University Press, 2009), 77.
6. Jimmy Carter, "The President's Proposed Energy Policy," April 18, 1977, http://www.pbs.org/wgbh/americanexperience/features/primary-resources/carter-energy.
7. For an excellent overview of policy measures for improving world energy markets, see Daniel P. Ahn, "Improving Energy Market Regulation: Domestic and International Issues," CGS/IIGG Working Paper, Council on Foreign Relations (February 2011). These policy recommendations are based on many of those presented in that paper as they relate to transparency in the physical and financial markets for oil.
8. Mary Fagan, "Sheikh Yamani Predicts Price Crash as Age of Oil Ends," *Telegraph*, June 25, 2000.

BIBLIOGRAPHY

Acheson, Edward G. "Why Coal and Oil Conservation." *The Forum*. May 1919.

Adelman, M. A. "Mineral Depletion, with Special Reference to Petroleum." *Review of Economics and Statistics* 72, no. 1 (February 1990): 1–10.

Adelman, M. A. "The World Oil Market: Past and Future." *The Energy Journal* 15, Special Issue (1994): 1–11.

Adelman, Morris A. *The Genie out of the Bottle: World Oil since 1970*. Cambridge, MA: MIT Press, 1995.

Adelman, M. A., and G. C. Watkins. "Reserve Asset Values and the Hotelling Valuation Principle: Further Evidence." *Southern Economic Journal* 61, no. 3 (January 1995): 664–673.

Advisory Commission to the Council of National Defense. *Defense* 2, no. 13. Official Weekly Bulletin of the Office of Emergency Management (March–April 1941).

Ahn, Daniel P. "Improving Energy Market Regulation: Domestic and International Issues." CGS/IIGG Working Paper. Council on Foreign Relations (February 2011).

Akerlof, George A., and Robert J. Shiller. *Animal Spirits: How Human Psychology Drives the Economy and Why it Matters for Global Capitalism*. Princeton, NJ: Princeton University Press, 2010.

Akin, James E. "The Oil Crisis: This Time the Wolf Is Here." *Foreign Affairs* 51, no. 3 (April 1973): 462–490.

American Petroleum Institute. *Petroleum Facts and Figures: Centennial Edition*. New York: American Petroleum Institute, 1959.

Andrews, Steve, and Randy Udall. "Peak Oil: It's the Flows, Stupid." Resilience.org, May 12, 2008. http://www.resilience.org.

"Arabian Oil Plan Defended by Knox." *New York Times*, March 22, 1944.

"Areas of Growth." Petroleum Press Service (July 1969).

Arlington Institute, The. "The Oil Depletion Analysis Centre." http://www.arlingtoninstitute.org/oil-depletion-analysis-centre.

Arnold, Ralph. "The Petroleum Resources of the United States." *Economic Geology* 10, no. 8 (December 1915): 695–712. Reprinted in *Annual Report of Board of Regents of the Smithsonian Institution Showing the Operations, Expenditures, and Condition of the Institution for the Year Ending June 30, 1916*. Washington, DC: Government Printing Office, 1917.

Bartlett, Albert A. "Sustained Availability: A Management Program for Non-renewable Resources." *American Journal of Physics* 54, no. 5 (May 1986): 398–402.

Bartlett, Albert A. "Reflections on Sustainability, Population Growth, and the Environment." *Population & Environment* 16, no. 1 (September 1994): 5–35.

Beckerman, Wilfred. "Economists, Scientists, and Environmental Catastrophe." *Oxford Economic Papers* 24, no. 3 (November 1972): 327–344.

Beckerman, Wilfred. *In Defence of Economic Growth*. London: J. Cape, 1974.

Benner, Katie. "Lawmakers: Will We Run Out of oil?" CNNMoney.com, December 7, 2005.

Bennett, Oliver. *Cultural Pessimism: Narratives of Decline in the Postmodern World*. Edinburgh: Edinburgh University Press, 2001.

Black, Brian. *Petrolia: The Landscape of America's First Oil Boom*. Baltimore, MD: Johns Hopkins University Press, 2000.

Bloch, Harry, and David Sapsford. "Innovation, Real Primary Commodity Prices and Business Cycles." Paper presented at the 13th conference of the International Joseph A. Schumpeter Society, Aalborg, Denmark, June 21–24, 2010.

Board of Governors of the Federal Reserve System. "Trade-Weighted U.S. Dollar Index." http://research.stlouisfed.org/.

Bowers, Tom. *Oil: Money, Politics, and Power in the 21st Century*. New York: Grand Central Publishing, 2010.

Bowman, Dylan. "Crude to Hit $175, Says Goldman Sachs." http://arabianbusiness.com, March 15, 2008.

BP. *BP Statistical Review of World Energy June 2010*. London, 2010. http://www.bp.com/statisticalreview.

BP. *BP Statistical Review of World Energy June 2011*. London, 2011. http://www.bp.com/statisticalreview.

BP. *BP Statistical Review of World Energy June 2012*. London, 2012. http://www.bp.com/statisticalreview.

Bradley, P. G., and G. C. Watkins. "Detecting Resource Scarcity: The Case of Petroleum." Proceedings of the IAEE 17th International Energy Conference, Stavanger, Norway, May 25–27. Volume 2, 1994.

Bradsher, Keith. "Vehicles to Vie for King of Hill." New York Times News Service. *Sarasota Herald Tribune*, June 17, 1997.

Brands, H. W. *T.R.: The Last Romantic*. New York: Basic Books, 1997.

Brinkley, Douglas. *Wheels for the World: Henry Ford, his Company, and a Century of Progress, 1903–2003*. New York: Penguin, 2003.

Campbell, Colin J. *The Golden Century of Oil 1950–2050: The Depletion of a Resource*. New York: Springer, 1991.

Campbell, Colin J. "The Twenty First Century: The World's Endowment of Conventional Oil and its Depletion." January 1996. http://www.oilcrisis.com/campbell/camfull.htm.

Campbell, Colin J. "The Coming Oil Crisis." *Quarterly Review of Economics and Finance* 42(1997): 373–389.

Campbell, Colin J. *Oil Crisis*. Essex: Multi-Science Publishing Co., 2005.

Campbell, Colin J. "About Peak Oil: Understanding Peak Oil." APSO International. http://www.peakoil.net/about-peak-oil.

Campbell, Colin J., and Jean H. Laherrère. "The End of Cheap Oil," *Scientific American*. March 1998.

"Careful Investigation Shows Gasoline Shortage Does Not Exist." *Los Angeles Times*, January 20, 1918.

Carnegie, Andrew. *The Autobiography of Andrew Carnegie and the Gospel of Wealth*. Digireads.com Publishing, 2009.

Carnegie, Andrew. *The Autobiography of Andrew Carnegie*. New York: PublicAffairs, 2011.

Carroll, Joe. "Global Oil Output Won't Peak until 2030, Yergin Says (Update1)." *Bloomberg*, November 14, 2006.

Carson, Rachel. *Silent Spring*. Boston: Houghton Mifflin, 1962.

Carter, Jimmy. "The President's Proposed Energy Policy." April 18, 1977. http://www.pbs.org/.

Cashin, Paul, and C. John McDermott. "The Long-Run Behavior of Commodity Prices: Small Trends and Big Variability." *IMF Staff Papers* 49, no. 2. International Monetary Fund, 2002.

Cashin, Paul, C. John McDermott, and Alasdair Scott. "Characteristics of the Current Commodity Boom." In *Global Economic Prospects 2009: Commodities at the Crossroads*. Washington, DC: The World Bank, 2009.

Cassidy, Padraic. "Goldman Sees Oil Price 'Super Spike.'" MarketWatch.com. *Wall Street Journal*, March 31, 2005.

"Charge of the Large." News Service Detroit, *New York Times*, June 26, 1997.

"The CIA Reassess the Geopolitics of Oil." Special Supplement. *Petroleum Intelligence Weekly*,
May 19, 1980.

Clayton, Blake. "Drilling into the American Energy Boom, in Four Charts." Energy, Security,
and Climate. CFR.org, January 8, 2013. http://blogs.cfr.org/levi/.

"Commodities: Crowded Out." *Economist.com*, September 24, 2011.

"Coolidge Appoints 4 Cabinet Members an Oil-Saving Board." *New York Times*, December 20,
1924.

CSP Daily News. "S.O.S. Poll: 67% of Americans Want New Oil Market Regulations." CSPnet.
com, July 17, 2008.

Cuddington, John T. "Calculating Long-Term Trends in the Real Prices of Primary Commodities: Deflator Adjustment and the Prebisch-Singer Hypothesis." Working Paper. August
29, 2007 (updated September 14, 2007).

Dahl, Erik J. "Naval Innovation: From Coal to Oil." *North American Shale Quarterly. E&P
Magazine* (July 4, 2006).

Daly, Herman E. "Steady-State Economics." In *Thinking about the Environment: Readings on
Politics, Property, and the Physical World*, edited by Matthew Alan Cahn and Rory O'Brien.
New York: M. E. Sharpe, 1996.

"Davies Depicts Gasoline Crisis." *New York Times*, August 29, 1941.

Day, David T. "The Petroleum Resources of the United States." In U.S. Geological Survey.
Papers on the Conservation of Mineral Resources. Bulletin 394. Washington, DC: Government Printing Office, 1909.

Day, David T. "The Petroleum Resources of the United States." In *The American Review of Reviews* 39 (January–June 1909). Edited by Albert Shaw.

Deffeyes, Kenneth S. *Hubbert's Peak: The Impending World Oil Shortage*. Princeton, NJ: Princeton University Press, 2001.

DeNovo, John A. "The Movement for an Aggressive American Oil Policy Abroad, 1918–1920."
American Historical Review 61, no. 4 (July 1956): 854–876.

Deslauriers, Michael J. "Famed Oil Tycoon Sounds Off on Peak Oil." *Resource Investor*, June 23,
2005.

Dvir, Eyal, and Kenneth S. Rogoff. "Three Epochs of Oil." NBER Working Paper No. 14927.
National Bureau of Economic Research: April 2009.

Dwoskin, Elizabeth. "Poll: Obama Dodges Blame for Gas Prices." *Bloomberg Businessweek*,
March 13, 2012.

"Earth Day: The History of a Movement." Earth Day Network. http://www.earthday.org/
earth-day-history-movement.

"Economist Commodity Price Index: Weights in the Index." *Economist*. http://media.
economist.com/media/pdf/Weights2005.pdf.

Edwards, John D. "Crude Oil and Alternate Energy Production Forecasts for the Twenty-First
Century: The End of the Hydrocarbon Era." *American Association of Petroleum Geologists
Bulletin* 81, no. 8 (1997): 1292–1305.

"Energy Risks and Energy Futures." *Wall Street Journal*, August 23, 1979.

"Epithets of the Week." *Life* 9, no. 10 (September, 2, 1940): 22.

"Executive Sees No Danger of Oil Shortage." *Christian Science Monitor*, July 17, 1947.

Exxon Corporation. *1979 Annual Report*. 1979.

Fagan, Mary. "Sheikh Yamani Predicts Price Crash as Age of Oil Ends." *Telegraph*, June 25,
2000.

Fanning, Leonard M., editor. *Our Oil Resources*. 2nd ed. New York: McGraw-Hill, 1950.

Farnsworth, Clyde H. "Trading Bloc Told Oil Dearth Looms." *New York Times*, May 26, 1972.

Fattouh, Bassam. "An Anatomy of the Crude Oil Pricing System." WPM 40. Oxford Institute for Energy Studies: January 2011.

Fattouh, Bassam, Lutz Kilian, and Lavan Mahadeva. "The Role of Speculation in Oil Markets: What Have We Learned So Far?" Working Paper (June 30, 2012).

"Fears of Petroleum Shortage Held Greatly Exaggerated." *Science News Letter*, August 31, 1935.

Frey, John W., and H. Chandler Ide. *A History of the Petroleum Administration for War, 1941–1945*. Washington, DC: United States Government Printing Office, 1946.

"Fuel Scarcity." *Wall Street Journal*, October 27, 1920.

"The Gasoline Anti-Climax." *Hartford Courant*, October 24, 1941.

"Gas, Fuel Oil Pinch Looms, Krug Says." *New York Times*, June 18, 1947.

"Gasoline Famine Panic Exploded by U.S. Reports." *Chicago Daily Tribune*, June 27, 1920.

"Gasoline for Farmers." *Indiana Farmer's Guide*, March 9, 1918.

"Gasoline Prices May Rise with Demand." *The Hartford Courant*, July 14, 1910.

"Gasoline Scarcity Rumor Questioned." *Christian Science Monitor*, September 18, 1918.

"Gasoline to Jump Seven Cents Gallon on First Day of Year." *Christian Science Monitor*, December 16, 1912.

"General Outlook for Oil." *Oil and Gas Journal*, October 10, 1912.

Georgescu-Roegen, Nicholas. *The Entropy Law and the Economic Process*. Cambridge, MA: Harvard University Press, 1971.

Georgescu-Roegen, Nicholas. "Energy and Economic Myths." *Southern Economic Journal* 41, no. 3 (January 1975): 247–381.

Gerth, Jeff. "Forecast of Rising Oil Demand Challenges Tired Saudi Fields." *New York Times*, February 24, 2004.

Gilbert, Chester G., and Joseph E. Pogue. "Petroleum: A Resource Interpretation." Bulletin 102, Part 6 PL. 1. Smithsonian Institution. United States National Museum. Washington, DC: Government Printing Office, 1918.

Glennie, K. W. *Introduction to the Petroleum Geology of the North Sea*. Oxford: Blackwell Scientific Publications, 1984.

"Global Reserves of Crude Oil Could Peak by Year 2010, Say Int'l Experts," Associated Press, May 26, 2002.

Goldman Sachs. "Underinvestment in Commodities Means Markets Will Be Tighter, Sooner." CEO Confidential, Issue 2002/05 (April 2002).

Google Books NGram Viewer. http://books.google.com/ngrams.

Google Zeitgeist. "Top of Mind," 2008, http://www.google.com/intl/en/press/zeitgeist2008/mind.html.

Gordon, Richard L. "IAEE Convention Speech: Energy, Exhaustion, Environmentalism, and Etatism." *Energy Journal* 15, no. 1 (1994): 1–16.

Gorelick, Steven M. *Oil Panic and the Global Crisis: Predictions and Myths*. Oxford: Blackwell, 2009.

Gray, John. *False Dawn*. New York: New Press, 1998.

Grilli, Enzo, and Maw, Cheng Yang. "Primary Commodity Prices, Manufactured Goods Prices, and the Terms of Trade of Developing Countries: What the Long Run Shows." *The World Bank Economic Review 2*, no. 1 (1988): 1–47.

Haber, Stephen, Noel Maurer, and Armando Razo. "When the Law Does Not Matter: The Rise and Decline of the Mexican Oil Industry." *Journal of Economic History* 63, no. 1 (March 2003): 1–31.

Hall, Charles A. S., and Carlos A. Ramirez-Pascualli. *The First Half of the Age of Oil: An Exploration of the Work of Colin Campbell and Jean Laherrère*, Springer Briefs in Energy, Energy Analysis. New York: Springer, 2013

"Hallanan Calls Crude Oil Price Rise Necessary to Avert Shortage." *Wall Street Journal*, May 10, 1943.

Hamilton, James D. "Understanding Crude Oil Prices." Working paper. May 22, 2008 (Revised December 6, 2008). http://dss.ucsd.edu/~jhamilto/understand_oil.pdf.

Hamilton, James D. "Cause and Consequences of the Oil Shock of 2007–08." Brookings Papers on Economic Activity. Brookings Institution (Spring 2009).

Hamilton, James D. "Oil Prices, Exhaustible Resources, and Economic Growth." In *Handbook on Energy and Climate Change*. Edited by Roger Fouquet. Cheltenham, U.K., and Northampton, MA: Edward Elgar Publishing, 2013.

Hargreaves, Steve. "Why $120 Oil Is Good." CNNMoney.com, May 8, 2008.

Harvey, David I., Neil M. Kellard, Jakob B. Madsen, and Mark E. Wohar. "The Prebisch-Singer Hypothesis: Four Centuries of Evidence." *Review of Economics and Statistics* 92, no. 2 (May 2010): 367–377.

Hays, Kristen. "Peak Oil Debate Crackles Anew." *Houston Chronicle*, November 15, 2006.

Hazlitt, A. J. "Tremendous Oil Possibilities of Great Northwest Texas Region." *Oil Trade Journal* 9 (February 1918).

Heap, Alan. *China—The Engine of a Commodities Super Cycle*. Smith Barney, Citigroup Global Markets, March 31, 2005.

Hiemstra, Glen. "Matt Simmons Sees $300 Oil." Futurist.com, http://www.futurist.com/2008/03/10/matt-simmons-sees-300-oil/.

Hirsch, Michael. "The Energy Wars." *Newsweek*, May 3, 2006.

Hirsch, Robert L., Roger Bezdek, and Robert Wendling. *Peaking of World Oil Production: Impacts, Mitigation, & Risk Management*. Report prepared by Science Applications International Corporation for the National Energy Technology Laboratory. U.S. Department of Energy: February 2005.

"Home Front: Oil or No Oil." *Time*, September 15, 1941.

Hotelling, Harold. "The Economics of Exhaustible Resources." *Journal of Political Economy*, 39, no. 2 (April 1931): 137–175.

"How Scarce is Oil." *Economist*, December 15, 1973.

Hubbert, M. K. "Nuclear Energy and the Fossil Fuels." *Drilling and Production Practice*, American Petroleum Institute (1956), http://www.mkinghubbert.com/files/1956.pdf.

Hubbert, M. K. "Nuclear Energy and the Fossil Fuels." Paper presented before the Spring Meeting of the Southern District, Division of Production, American Petroleum Institute, San Antonio, Texas, March 7–9, 1956 (Publication NO. 95, Shell Development Company Exploration and Production Research Division, Houston, Texas, June 1956), http://www.mkinghubbert.com/files/1956.pdf.

Ibrahim, Youssef M. "Falling Oil Prices Pinch Several Producing Nations." *New York Times*, June 23, 1998.

"Ickes 'Gas' Shortage Scare Due to Anxiety over Future." *Christian Science Monitor*, September 19, 1941.

"Ickes Urges National D.S.T.; Power, Oil Shortage Foreseen." *Christian Science Monitor*, May 31, 1941.

International Energy Agency. *World Energy Outlook 1998*. Paris: IEA Publications, 1998.

International Energy Agency. *World Energy Outlook 2000*. Paris: IEA Publications, 2000.

International Energy Agency. *Oil Market Report: 11 June 2000*. 2000. http://www.oilmarketreport.org/.

International Energy Agency. *Oil Market Report: 11 December 2000*. 2000. http://www.oilmarketreport.org/.

International Energy Agency, *Oil Market Report: 11 September 2002*. 2002. http://www.oilmarketreport.org/.

International Energy Agency. *Oil Market Report: 11 December 2002*. 2002. http://www.oilmarketreport.org/.

International Energy Agency. *Oil Market Report: 15 March 2002*. 2002. http://www.oilmarketreport.org/.

International Energy Agency. *Oil Market Report: 17 January 2003*. 2003. http://www.oilmarketreport.org/.

International Energy Agency. *Oil Market Report: 12 March 2003*. 2003. http://www.oilmarketreport.org/.

International Energy Agency. *Oil Monthly Report: 10 October 2003.* 2003. http://www.oilmarketreport.org/.

International Energy Agency. *Oil Market Report: 11 October 2007.* 2007. http://www.oilmarketreport.org/.

International Energy Agency. *Oil Market Report: 12 December 2012.* 2012. http://www.oilmarketreport.org.

International Monetary Fund. World Economic Outlook Database. April 2013. http://www.imf.org/.

International Energy Agency. "Supply Shock from North American Oil Rippling through Global Markets." Press Release, May 14, 2013. http://www.iea.org/.

Investigation of the National Defense Program. *Additional Report: Report of Subcommittee Concerning Investigations Overseas.* Report 10, Part 15, Senate, 78th Congress, Second session. Washington, DC: Government Printing Office, 1944. Reprinted in Investigation of the National Defense Program. *Additional Report of the Special Committee Investigating the National Defense Program.* Report No. 10, Part 16, Senate, 78th Congress, Second session. Washington, DC: Government Printing Office, 1944.

Isser, Steve. *The Economics and Politics of the United States Oil Industry, 1920–1990: Profits, Populism, and Petroleum.* New York: Garland Publishing, Inc., 1996.

Ivanhoe, L. F. "Get Ready for Another Oil Shock!" *The Futurist* 31, no. 1 (January-February 1997).

Jaffe, Amy Myers. "Not So Cheap." *Foreign Affairs*, May 26, 2004.

Jaffe, Amy Myers. "The Americas, Not the Middle East, Will Be the World Capital of Energy." *Foreign Policy*, no. 188 (September/October 2011), 86–87.

Jaffe, Amy Myers, and Robert A. Manning, "The Shocks of a World of Cheap Oil." *Foreign Affairs* 79, no. 1 (January/February 2000), 16–29.

Jense, Niels. "The Case for Human Ingenuity." *Credit Writedowns*, May 3, 2011. http://www.creditwritedowns.com/.

Jevons, William Stanley. *The Coal Question.* 2nd ed. London: Macmillan and Co, 1866.

Johnson, Kir. "From a Woodland Elegy, A Rhapsody in Green; Hunter Mountain Paintings Spurred Recovery." *New York Times*, June 7, 2001. http://www.nytimes.com/.

Kagan, Robert. *The World America Made.* New York: Vintage Books, 2013.

Kissinger, Henry. *Years of Upheaval: The Second Volume of His Classic Memoirs.* New York: Simon & Shuster Paperbacks, 2011.

Klare, Michael T. "The Saudi Oil Bombshell." Asia Times Online, June 29, 2005.

Klare, Michael T. "Tomgram: Michael Klare, Is Washington Out of Gas?," TomDispatch.com, September 15, 2011. http://www.tomdispatch.com/archive/175441/.

Kline, Benjamin. *First along the River: A Brief History of the U.S. Environmental Movement.* 3rd ed. Lanham, MD: Rowman & Littlefield Publishers, Inc., 2007.

Knowlton, Brian. "Gore Urges Use of Oil Reserves to Ease Prices." *New York Times*, September 22, 2000.

Krauss, Clifford. "There Will be Fuel." *New York Times*, November 16, 2010.

Krauss, Clifford. "Shale Boom in Texas Could Increase U.S. Oil Output." *New York Times*, May 27, 2011.

Kronenberg, Tobias. "Should We Worry about the Failure of the Hotelling Rule?" Paper prepared for Monte Verità Conference on Sustainable Resource Use and Economic Dynamics, June 4–9, 2006, Ascona, Switzerland. Center of Economic Research at ETH Zurich.

"Krug Tells Compact Substantial Oil Imports will be Needed." *Oil and Gas Journal*, August 17, 1946.

"Krug Warns Oil Shortage Here to Stay." *Christian Science Monitor*, February 16, 1948.

Kuykendall, Chris. "M. King Hubbert and His Successors: A Half-Bibliography through 2005." March 2006. http://www.mkinghubbert.com/files/HubbertBibliography.20060308.pdf.

"Large Oil Supply Seen." *New York Times*, May 7, 1948.

Levi, Michael. *The Power Surge: Energy, Opportunity, and the Battle for America's Future.* New York: Oxford University Press, 2013.

Levy, Walter J. "Oil Power." *Foreign Affairs,* July 1971.

Lynch, Michael C. "Crying Wolf: Warnings about Oil Supply." March 1998. http://sepwww.stanford.edu/sep/jon/world-oil.dir/lynch/worldoil.html.

Lynch, Michael C. "The New Pessimism about Petroleum Resources: Debunking the Hubbert Model (and Hubbert Modelers)." *Minerals & Energy—Raw Materials Report* 18, no. 1 (2003), 21–32.

Maguire, A. G. *Prices and Marketing Practices Covering the Distribution of Gasoline and Kerosene throughout the United States.* U.S Fuel Administration, Oil Division. Washington, DC: Government Printing Office, 1919.

Malthus, Thomas Robert. *An Essay on the Principle of Population.* 6th ed. London: John Murray, 1826.

Mann, Charles C. "What If We Never Run Out of Oil?" *The Atlantic,* April 24, 2013.

"Markets & Data." *Economist.* http://www.economist.com/.

Maugeri, Leonardo. *The Age of Oil: The Mythology, History, and Future of the World's Most Controversial Resource.* Westport, CT: Praeger, 2006.

"Mayor Warns of Probable Fuel Oil Shortage." *Hartford Courant,* June 4, 1941.

McCabe, P. J. "Energy Resources—Cornucopia or Empty Barrel?" *American Association of Petroleum Geologists Bulletin* 11 (1998): 2110–2134.

McCall, William. "Physicist Warns Global Oil Production Peak Could Bring Economic Disaster." Associated Press, August 13, 2001.

McNulty, Sheila. "Oil Back in Favor with US Drillers after Years of Targeting Gas." *Energy Source. Financial Times,* July 14, 2010. http://blogs.ft.com/energy-source/2010/07/14/oil-back-in-favor-with-us-drillers-after-years-of-targeting-gas/#axzz2ZFR6ZkAf.

Meadows, Donella H., Dennis L. Meadows, Jorgen Randers, and William W. Behrens III. *Limits to Growth.* New York: Universe Books, 1972.

Meadows, Donella H., et al. *Limits to Growth: A Report for the Club of Rome's Project on the Predicament of Mankind.* 2nd ed. New York: Universe Publishing, 1974.

Mejcher, Helmut. *Imperial Quest for Oil: Iraq 1910–1928.* Middle East Centre. St. Antony's College. Oxford, UK: Ithaca Press, 1976.

Melby, Eric D. K. *Oil and the International System: The Case of France, 1918–1969.* New York: Arno Press, 1981.

"Merger Won't Fix Oil-Glut Fallout." *The Florida Times-Union,* December 2, 1998.

Merline, John. "Public Misplaces Blame for High Oil Prices." *Investor's Business Daily,* May 11, 2011.

"Military Oil Needs will Increase; Civilians to Get Less—Ickes." *National Petroleum News,* February 24, 1943.

Mill, John Stuart. *Principles of Political Economy: With Some of Their Applications to Social Philosophy.* 5th ed. London: Parker, Son, and Bourn, 1862.

Miller, Char. *Gifford Pinchot and the Making of Modern Environmentalism.* Washington, DC: Island Press, 2001.

Mintz, John. "Outcome Grim at Oil War Game." *Washington Post,* June 24, 2005.

Mishan, Ezra J. "Growth and Antigrowth: What Are the Issues?" In *The Economic Growth Controversy,* edited by Andrew Weintraub, Eli Schwartz, and Jay Richard Aronson. London: Macmillan, 1974.

Mitchell, Alison. "Senator Promises Investigation into Gasoline Price Rise." *New York Times,* May 30, 2001.

"Model T Ford Production." Compiled by R. E. Houston, Ford Production Department. August 3, 1927. http://www.mtfca.com/encyclo/fdprod.htm.

"Moffett Hints U.S. Control of Oil Motivates Claims Shortage Exists." *Wall Street Journal,* October 4, 1941.

Montague, Gilbert H. *The rise and progress of the Standard Oil Company.* New York: Harper and Brothers, 1903.

Morse, Edward, and Michael Waldron, "Oil Dot-com." Lehman Brothers Energy Special Report, May 29, 2008.

Morse, Edward L., et al. *Energy 2020: Independence Day.* Citi GPS: Global Perspectives and Solutions, February 2013.

Mouawad, Jad. "With Oil Prices Off Their Peak, Are Supplies Assured?" *New York Times*, December 5, 2005.

Mouawad, Jad. "The Big Thirst." *New York Times*, April 20, 2008.

Mouawad, Jad. "Oil Price Rise Fails to Open Tap." *New York Times*, April 29, 2008.

Mouawad, Jad. "Big Oil Projects Put in Jeopardy by Fall in Prices." *New York Times*, December 15, 2008.

Murrell, John H. "Middle East Oil." *Tulsa Geological Society Digest* 14 (1945–1946).

Nasaw, David. *Andrew Carnegie.* New York: Penguin Press, 2006.

Nash, Roderick Frazier. *American Environmentalism: Readings in Conservation History.* 3rd ed. New York: McGraw-Hill, 1990.

"Nation Faces Oil Famine." *Los Angeles Times*, September 23, 1923.

"Nation Warned It Faces Oil and Gas Shortage." *Chicago Daily Tribune*, June 18, 1947.

"A National Report on America's Energy Crisis." Remarks by U.S. Secretary of Energy Spencer Abraham at the U.S. Chamber of Commerce. National Energy Summit, March 19, 2001.

Neumayer, Eric. "Scarce or Abundant? The Economics of Natural Resource Availability." *Journal of Economic Surveys* 14, no. 3 (2000), 307–335.

Newport, Frank. "Americans Rate Computer Industry Best, Oil and Gas Worst." Gallup Economy, August 16, 2012. http://www.gallup.com/poll/156713/americans-rate-computer-industry-best-oil-gas-worst.aspx.

Nixon, Richard. "Address to the Nation about Policies to Deal with the Energy Shortages." November 7, 1973.

Nocera, Joseph. "On Oil Supply, Opinions Aren't Scarce." *New York Times*, September 10, 2005.

Nooten, George A. "Sustainable Development and Nonrenewable Resources—A Multilateral Perspective." *Proceedings of the Workshop on Deposit Modeling, Mineral Resource Assessment, and Sustainable Development.* 2007.

Nordhaus, William D. "Resources as a Constraint on Growth." *American Economic Review* 64, no. 2 (May 1974): 22–26.

O'Connor, John, and David Orsmond. "The Recent Rise in Commodity Prices: A Long-Run Perspective," *Reserve Bank of Australia Bulletin* (April 2007).

"Oil Burning in the Ships." *Oil and Gas Journal* 17, no. 29 (December 20, 1918).

"Oil: Cold Comfort." *Time*, February 2, 1948.

"Oil Companies Discount Reports of Gas Shortage." *Christian Science Monitor*, October 14, 1947.

"Oil Curbs Seen Prolonging Shortage." *Christian Science Monitor*, January 29, 1948.

"Oil Demand Is Exceeding the Rate of Production." *Christian Science Monitor*, March 14, 1913.

"Oil Depletion: Sunset for the Oil Business?" Economist, November 1, 2001.

"Oil Famine Alarmists." *Los Angeles Times*, March 6, 1926.

"Oil: Famine Closer." *Time* 38, no. 4 (July 28, 1941).

"Oil Men Face Dilemma in Meeting Shortages." *Christian Science Monitor*, February 25, 1948.

"Oil 'Peak' Not Seen Coming Any Time Soon." Associated Press, June 21, 2005.

"Oil: Petroleum Economy." *Time*, February 16, 1948.

"Oil Shocked." *Economist*, March 26, 1998.

"Oil Shortage has Arrived, U.S. Aid Warns." *Chicago Tribune*, August 21, 1941.

"Oil Shortage Seen for United States." *Christian Science Monitor*, March 15, 1918.

"Oil Shortages in the East Phony, Says Rep. Fish." *Chicago Daily Tribune*, August 26, 1941.

"Oil: Summer Shortage." *Time* 50, no. 1 (July 7, 1947).

"Oil: Quick Change." *Time* 56, no. 4 (July 24, 1950).

O'Keefe, Brian. "Exxon's Big Bet on Shale Gas." CNNMoney.com, April 16, 2012.

Olien, Diana Davids, and Roger M. Olien. "Running Out of Oil: Discourse and Public Policy 1909–1929." *Business and Economic History* 22, no. 2 (Winter 1993): 36–66.

"OPEC Nations Seen Lifting Ceiling on Oil Production." *New York Times*, November 27, 1997.

"Out of Gas?" *Time* 42, no. 1 (July 5, 1943).

Ovide, Shira. "A Career 'Super Spike' for Goldman's Infamous Oil Analyst." Deal Journal. *Wall Street Journal*, February 1, 2011. http://blogs.wsj.com/deals/.

Painter, David S. *Oil and the American Century: The Political Economy of US Foreign Oil Policy, 1941–1954.* Baltimore: Johns Hopkins University Press, 1986.

Parra, Francisco. *Oil Politics: A Modern History of Petroleum.* New York: I. B. Taurus and Co., 2010.

Patalon III, William. "Goldman Sachs Follows Money Morning Prediction That Oil Prices Could Approach $200 a Barrel." Money Morning, March 17, 2008. http://moneymorning.com/.

"Paulson: 'No Quick Fix' to Oil Prices." *USAToday*, June 1, 2008.

Petroleum Intelligence Weekly. August 13, 1973.

Petroleum Intelligence Weekly. September 10, 1973.

Pew Research Center for the People & the Press. "From News Interest to Lifestyles, Energy Takes Hold." May 24, 2001.

"Pinchot is for Conservation." *Oil and Gas Journal*, March 2, 1916.

"Plymouth Oil President Warns of Crude Shortages." *Wall Street Journal*, March 26, 1943.

"Pointing to our Need for Foreign Oil Supply Places Industry at Fork of Road to Control." *National Petroleum News*, August 21, 1946.

Prebisch, Raúl. *The Economic Development of Latin America and Its Principal Problems.* New York: United Nations Department of Economic Affairs, 1950.

"Prebisch-Singer Hypothesis." *International Encyclopedia of the Social Sciences* (2008).

"Present Gasoline Supply Ample for All." *Motor Age*, July 18, 1918.

"President Orders Federal Heat Cut and Autos Slowed." *New York Times*, January 18, 1948.

"Probe Sought in Bay State Oil Shortage." *Christian Science Monitor*, January 13, 1948.

Radetzki, Marian. *A Handbook of Primary Commodities in the Global Economy.* Cambridge, MA: Cambridge University Press, 2008.

Radetzki, Marian. *Handbook of Primary Commodities in the Global Economy.* Reissue edition. New York: Cambridge University Press, 2010.

Reinhart, Carmen M., and Kenneth S. Rogoff. *This Time is Different: Eight Centuries of Financial Folly.* Princeton, NJ: Princeton University Press, 2009.

Report of the National Conservation Commission. Volume I. Senate Document No. 676, 60th Congress, Second session. Washington, DC: Government Printing Office, 1909.

Requa, M. L. "An Article on the Exhaustion of the Petroleum Resources of the United States, Showing the Present and Future Supply and Demand, Also the Production of the Principal Oil Fields of the United States." *Petroleum Resources of the United States.* U.S. Senate Document No. 363, 64th Congress, First session. Washington, DC: Government Printing Office, 1916.

"Responses to Daniel Yergin's Attack on Peak Oil." www.resilience.org. September 19, 2011.

Ricardo, David. *On the Principles of Political Economy and Taxation.* 3rd ed. London: John Murray, 1821.

Robinson, Albert G. "The Fuel of the Future." *Outlook.* July 21, 1920.

Rockefeller Foundation. *Working Paper on International Energy Supply: An Industrial World Perspective.* New York: Rockefeller Foundation, 1978.

"Roosevelt Links Pipeline to Needs." *New York Times*, March 4, 1944.

Roosevelt, Theodore. "Seventh Annual Message to Congress." December 3, 1907. http://www.pbs.org/weta/thewest/resources/archives/eight/trconserv.htm.

Salpukas, Agis. "Oil's Numbers Game." *New York Times*, December 1, 1997.

Salpukas, Agis. "Challenges Inside and Out Confront OPEC." *New York Times*, June 23, 1998.

Sanger, David E. "Singing the Cartel Blues." *New York Times*, March 29, 1998.

"Save Coal to Meet Industry's Greatest Peace Demand." *Power* 48, no. 22 (November 26, 1918).

"Says Oil Men Seek to Frighten Buyers." *New York Times,* June 15, 1941.

"Scarcity of Gas Oil Threatened." *Los Angeles Times,* December 27, 1914.

Scherer, Rob. "High Oil Prices Siphon Profits from US Firms." *Christian Science Monitor,* September 21, 2000.

Schlesinger, James R. "Energy Risks and Energy Futures: Some Farewell Observations." Supplement. *Petroleum Intelligence Weekly,* August 27, 1979.

"Schlesinger Warns West U.S. Energy Plan Is Vital; Program to Ease Reliance on Imported Oil Adopted by Ministers of I.E.A." *New York Times,* October 6, 1977.

Scott, Amanda. "Obama Talks Energy at the State of the Union 2013." Energy.Gov. http://energy.gov/articles/president-obama-talks-energy-state-union-2013.

"Self-Sufficiency Stressed in Senate Oil Policy Report." *National Petroleum News,* February 5, 1947.

"Senate Committee Finds No Gasoline Shortage." *Los Angeles Times,* September 12, 1941.

"Serious Petroleum Shortage Predicted in Near Future." *Science News Letter,* August 31, 1935.

SF Informatics. "Technology Won't Solve America's Oil Addiction, Experts Say." Press Release. PRWeb, February 2, 2006. http://www.prweb.com/releases/200614/2/prweb341058.htm.

"Shah of Iran Explains Latest Oil Price Increase." Supplement. *Middle East Economic Survey* 17 no. 10 (December 28, 1973).

Shenk, Mark. "Oil Jumps to Record $58.60 as Demand May Outpace Production." *Bloomberg,* June 17, 2005.

Shiller, Robert. *Irrational Exuberance.* 1st and 2nd ed. Princeton: Princeton University Press, 2000 and 2005.

Shiller, Robert. *Irrational Exuberance,* 2nd ed. (paperback). New York: Crown Business, 2006.

Shiller. Robert J. "Online Data Robert Shiller," Homepage. http://www.econ.yale.edu/~shiller/data.htm.

"A Shortage, But Not of Gas." *Chicago Tribune,* August 25, 1941.

"Shortage Expected to Last Five Years." *Christian Science Monitor,* February 20, 1948.

"Shortages of Oil Disappear—Stocks on Hand Accumulate." *Christian Science Monitor,* January 3, 1949.

Simmons, Matthew R. "Revisiting the Limits to Growth: Could the Club of Rome Have Been Correct, After All?" An Energy White Paper. Simmons & Company International (October 2000). Available on Simmons & Company website.

Simmons, Matthew R. *Twilight in the Desert: The Coming Saudi Oil Shock and the World Economy.* Hoboken, NJ: Wiley, 2005.

Simon, Julian L. "Resources, Population, Environment: An Over-supply of False Bad News." *Science* 208, no. 4451 (June 1980): 1431–1437.

Simon, Julian L. "False Bad News Is Truly Bad News." *The Public Interest,* no. 65 (1981): 80–90.

Singer, Hans. "The Distribution of Gains between Investing and Borrowing Countries." *American Economic Review* 40, no. 2 (1950).

Skeet, Ian. *OPEC: Twenty-Five Years of Prices and Politics.* Cambridge, UK: Cambridge University Press, 1988.

Smil, Vaclav. *Energy at the Crossroads.* Cambridge, MA: MIT Press, 2003.

Smith, G.O. "Our Mineral Resources." *Annals of the American Academy of Political and Social Science.* Philadelphia: American Academy of Political and Social Science, 1909.

Smith, James L. "World Oil: Market or Mayhem?" *Journal of Economic Perspectives* 23, no. 3 (Summer 2009): 145–164.

"Standard oil (Ind.) Rationing Gas in Midwest." *Chicago Tribune,* June 25, 1947.

Steigerwald, Bill. "Pickens: We can't Become Energy Independent." *Pittsburgh Tribune-Review,* February 12, 2006.

Stern, Roger. "Oil Scarcity Ideology in US National Security Policy, 1909–1980." Working Paper on the Oil, Energy & the Middle East Program. Princeton University, 2012.

Stevens, Paul J. "The Future Price of Crude Oil." *Middle East Economic Survey* 47, no. 37 (September 2004), http://web.archive.org/web/20041216111616/http://www.mees.com/postedarticles/oped/a47n37d01.htm.

Stevens, Paul J. "Oil Markets." *Oxford Review of Economic Policy* 21, no. 1 (2005): 19–42.

Stiglitz, Joseph. "Growth with Exhaustible Natural Resources: Efficient and Optimal Growth Paths." *Review of Economic Studies* 41, Symposium on the Economics of Exhaustible Resources (1974): 123–137.

Strauss, Lawrence C. "What Mr. Crude Oil Sees Ahead." *Barron's*, June 9, 2008. http://online.barrons.com/.

Sullivan, Colin. "Has 'Peak Oil' Gone the Way of the Flat Earth Society." EnergyWire, March 22, 2013.

"Supply Picture Unchanged Despite 'Scare' Headlines." *National Petroleum News*, June 25, 1947.

Swartz, Mimi. "The Gospel According to Matthew." *Texas Monthly*. February 2008.

Taylor, Jerry and Peter Van Doren. "An Oil Embargo Won't Work." *Wall Street Journal*, April 10, 2002.

Tilton, John E. *On Borrowed Time?: Assessing the Threat of Mineral Depletion.* Resources for the Future, 2003.

Tower. Samuel A. "Oil Running Short, House Body Warns." *New York Times*, May 7, 1948.

Trembath, Alex, Jesse Jenkins, Ted Nordhaus, and Michael Shellenberger. *Where the Shale Gas Revolution Came From: Government's Role in the Development of Hydraulic Fracturing in Shale.* Breakthrough Institute, May 2012. http://thebreakthrough.org/.

"U.S. and Europe Face Oil Shortage." *New York Times*, November 16, 1947.

"U.S. as an Oil-Importing Nation is Forecast by Technologist." *New York Times*, July 15, 1943.

U.S. Bureau of Labor Statistics. "Consumer Price Index." http://www.bls.gov/cpi/.

U.S. Central Intelligence Agency. *The International Energy Situation: Outlook to 1985.* Washington, DC: Government Printing Office, 1977.

U.S. Central Intelligence Agency. *The World Oil Market in the Years Ahead.* Washington, DC: Government Printing Office, 1979.

U.S. Department of State. "The International Oil Industry through 1980." December 1971. In Muslim Students Following the Line of the Imam. *Documents from the U.S. Espionage Den*, volume 57. Tehran: Center for the Publication of the U.S. Espionage Den's Documents, 1986.

U.S. Energy Information Administration, "Real Prices Viewer." http://www.eia.gov.

U.S. Energy Information Administration. "International Energy Statistics." http://www.eia.gov.

U.S. Energy Information Administration. "Long Term World Oil Supply: A Resource Base/Production Path Analysis." Presentation by Jay Hakes to the American Association of Petroleum Geologists on April 18, 2000, New Orleans, Louisiana. http://www.eia.gov/pub/oil_gas/petroleum/presentations/2000/long_term_supply/sld001.htm.

U.S. Energy Information Administration. "EIA Tracks U.S. Tight Oil Production as Volumes Soar." This Week in Petroleum, Release Date March 14, 2012.

U.S. Energy Information Administration. "Table 4.2 Crude Oil and Natural Gas Cumulative Production and Proved Reserves, 1977–2010." Annual Energy Review. Release date: September 27, 2012.

U.S. Energy Information Administration. "U.S. Ending Stocks Excluding SPR of Crude Oil." Release date March 15, 2013.

U.S. Energy Information Administration. "U.S. Field Production of Crude Oil." Release date: March 15, 2013.

U.S. Energy Information Administration. "U.S. Ending Stocks of Total Gasoline (Thousand Barrels)." Release date: May 30, 2013.

U.S. Energy Information Administration. "U.S. Ending Stocks of Distillate Fuel Oil (Thousand Barrels)." Release date: May 30, 2013.

U.S. Energy Information Administration. *Short-Term Energy Outlook*. Release Date June 11, 2013.

U.S. Federal Trade Commission. *Report of the Federal Trade Commission on the Pacific Coast Petroleum Industry: Part II Prices and Competitive Conditions*. Washington, DC: Government Printing Office, 1922.

U.S. Forest Service. "Gifford Pinchot (1865–1946)." Historical Information. http://www.fs.fed.us/gt/local-links/historical-info/gifford/gifford.shtml.

U.S. Geological Survey. *An Estimate of Undiscovered Conventional Oil and Gas Resources of the World 2012*. Fact Sheet 2012–2013 (March 2012). http://pubs.usgs.gov/fs/2012/3042/fs2012-3042.pdf.

U.S. Geological Survey World Assessment Team. "U.S. Geological Survey World Petroleum Assessment 2000—Description and Results." U.S. Geological Survey Digital Data Series—DDS-60. U.S. Department of the Interior. http://pubs.usgs.gov/dds/dds-060/.

U.S. House of Representatives. *Hearings before a subcommittee of the Committee on Interstate and Foreign Commerce*. House of Representatives, 77th Congress, Second session. Washington, DC: U.S. Government Printing Office, 1942.

U.S. House of Representatives. Hearings before the Committee on Foreign Affairs on H.R. 2362 A Bill to Amend the Economic Cooperation Act of 1948, Part 1. 81st Congress, First session. Washington, DC: United States Government Printing Office, 1949.

"U.S. Official Warns of Imports Danger." *Oil and Gas Journal*, May 15, 1972, 50.

U.S. Senate Committee on Energy and National Resources. "Part I—The Gathering Energy Crisis." *The Geopolitics of Oil*. Staff Report, 96th Congress, Second session. Washington, Government Printing Office, December 1980.

U.S. Senate. *Report of the Senate Select Committee on Interstate Commerce*. Senate Report 46 Part 1, 49th Congress, First session. Washington, DC: Government Printing Office, 1886.

U.S. Senate. Hearing before the Committee on Foreign Relations on the Nomination of Dean G. Acheson to be Secretary of State. 81st Congress, First session. Washington, DC: United States Government Printing Office, 1949.

U.S. Tariff Commission. *Petroleum: Prepared in Response to Requests from the Committee on Finance of the United States and the Committee on Ways and Means of the House of Representatives*. War Changes in Industry Series No. 17. Washington, DC: United States Government Printing Office, 1946.

"U.S. Urges Allied Unity on Oil Crisis." *The Spokesman-Review*, December 13, 1973.

"US View of OPEC-West Relations." *Petroleum Economist*, December 1979.

Vidal, John. "Analyst Fears Global Oil Crisis in Three Years." *Guardian*, April 26, 2005.

Voss, Stephen. "Saudi, U.S. Officials Confident Technology will Boost Reserves." *Bloomberg*, September, 13 2006.

"War Hits Nearer while Oil Fuel Famine Impends." *The Hartford Courant*, May 22, 1918.

Weintraub, Bernard. "Iran Keeps Oil Flowing Despite Reported Pressure from Arabs." *New York Times*, December 18, 1973.

"When the Wells Run Dry." *Oil and Gas Journal*, February 24, 1916.

White, David. "The Petroleum Resources of the World." *The Annals of the American Academy of Political and Social Science* 89 Prices (May 1920): 111–134.

Wildavsky, Aaron, and Ellen Tenenbaum. *The Politics of Mistrust: Estimating American Oil and Gas Resources*. Beverly Hills, CA: Sage Publications, 1981.

Williamson, Harold F. "Prophecies of Scarcity or Exhaustion of Natural Resources in the United States." *American Economic Review* 35, no. 2 (May 1945): 97–109.

Williamson, Harold F., Ralph L. Andreano, Arnold R. Daum, and Gilbert C. Klose. *The American Petroleum Industry: 1899–1959, the Age of Energy*. Evanston, IL: Northwestern University Press, 1963.

WorldPublicOpinion.org. "World Publics Say Oil Needs to Be Replaced as Energy Source."
    Press Release. April 20, 2008. http://www.worldpublicopinion.org/.
Yamni, Sheikh Ahmed Zaki. "Debate at the Oxford Energy Seminar, 13 September 1985." In
    OPEC and the World Market: The Genesis of the 1986 Price Crisis. Edited by Robert Mabro.
    Oxford: Oxford University Press—Oxford Institute for Energy Studies, 1986.
"300-Year Oil Supply Believed in America." New York Times, September 26, 1943.
Yergin, Daniel. The Prize: The Epic Quest for Oil, Money, and Power. New York: Simon & Schus-
    ter, 1992.
Yergin, Daniel. "It's Not the End of the Oil Age." Washington Post, July 31, 2005.
Yergin, Daniel. The Quest: Energy, Security, and the Remaking of the Modern World. New York:
    Penguin Press, 2011.
Yergin, Daniel. "There Will Be Oil." Wall Street Journal, September 27, 2011.
Yergin, Daniel, and Joseph Stanislaw. "How OPEC Lost Control of Oil." Time 151, no. 13
    (April 6, 1998).
Zabarenko, Deborah. "Oil Running Out as Prime Energy Source: World Poll." Reuters, April
    20, 2008.
Zook, Ralph. The Proposed Arabian Pipeline: A Threat to Our National Security. Tulsa, OK: In-
    dependent Petroleum Association of America, 1944.

INDEX

Note: Figures are indicated by "f" following the page number. Locators followed by the letter 'n' refer to notes.